An Illustrated Review of the DIGESTIVE SYSTEM

Glenn F. Bastian

HarperCollins*CollegePublishe*

030790Z

Executive Editor: Bonnie Roesch
Cover Designer: Kay Petronio
Production Manager: Bob Cooper
Printer and Binder: Malloy Lithographing, Inc.
Cover Printer: The Lehigh Press, Inc.

AN ILLUSTRATED REVIEW OF THE DIGESTIVE SYSTEM

by Glenn F. Bastian

Copyright © 1994 HarperCollins College Publishers

Library of Congress Cataloging-in-Publication Data
Bastian, Glenn F.
        An illustrated review of the digestive system / Glenn F. Bastian.
            p.   cm.
        Includes bibliographical references.
        ISBN: 0-06-501710-2
        1. Digestive organs—Physiology 2. Digestion. I. Title.
        [DNLM: 1. Digestion—physiology—programmed instruction.
2. Digestion—physiology—atlases. 3. Digestive System—anatomy &
histology—programmed instruction. 4. Digestive System—physiology—atlases.
WI 18 B326i 1994]
QP145.B344   1994
612.3'076—dc20
DNLM/DLC
for Library of Congress   94–1897
                                                                    CIP

95 96 9 8 7 6 5 4 3 2

To
Rebecca and Elwin Sykes

CONTENTS

LIST OF TOPICS & ILLUSTRATIONS

Text: One page of text is devoted to each of the following topics. *Illustrations are listed in italics.*

PREFACE

An Illustrated Review of Anatomy and Physiology is a series of ten books written to help students effectively review the structure and function of the human body. Each book in the series is devoted to a different body system.

My objective in writing these books is to make very complex subjects accessible and nonthreatening by presenting material in manageable size bits (one topic per page) with clear, simple illustrations to assist the many students who are primarily visual learners. Designed to supplement established texts, they may be used as a student aid to jog the memory, to quickly recall the essentials of each major topic, and to practice naming structures in preparation for exams.

INNOVATIVE FEATURES OF THE BOOK

(1) Each major topic is confined to one page of text.

A unique feature of this book is that each topic is confined to one page and the material is presented in outline form with the key terms in boldface or italic typeface. This makes it easy to scan quickly the major points of any given topic. The student can easily get an overview of the topic and then zero in on a particular point that needs clarification.

(2) Each page of text has an illustration on the facing page.

Because each page of text has its illustration on the facing page, there is no need to flip through the book looking for the illustration that is referred to in the text ("see Figure X on page xx"). The purpose of the illustration is to clarify a central idea discussed in the text. The images are simple and clear, the lines are bold, and the labels are in a large type. Each illustration deals with a well-defined concept, allowing for a more focused study.

PHYSIOLOGY TOPICS (1 text page : 1 illustration)
Each main topic in physiology is limited to one page of text with one supporting illustration on the facing page.

ANATOMY TOPICS (1 text page : several illustrations)
For complex anatomical structures a good illustration is more valuable than words. So, for topics dealing with anatomy, there are often several illustrations for one text topic.

(3) Unlabeled illustrations have been included.

In Part II, all illustrations have been repeated without their labels. This allows a student to test his or her visual knowledge of the basic concepts.

(4) A Pronunciation Guide has been included.

Phonetic spellings of unfamiliar terms are listed in a separate section, unlike other textbooks where they are usually found in the glossary or spread throughout the text. The student may use this guide for pronunciation drill or as a quick review of basic vocabulary.

(5) A glossary has been included.

Most textbooks have glossaries that include terms for all of the systems of the body. It is convenient to have all of the key terms for one system in a single glossary.

ACKNOWLEDGMENTS

I would like to thank the reviewers of the manuscript for this book who carefully critiqued the text and illustrations for their effectiveness: William Kleinelp, Middlesex County College; Robert Smith, University of Missouri, St. Louis, and St. Louis Community College, Forest Park; and Pamela Monaco, Molloy College. Their help and advice are greatly appreciated. Kay Petronio is to be commended for her handsome cover design and Bob Cooper has my gratitude for keeping the production moving smoothly. Finally, I am greatly indebted to my editor Bonnie Roesch for her willingness to try a new idea, and for her support throughout this project. I invite students and instructors to send any comments and suggestions for enhancements or changes to this book to me, in care of HarperCollins, so that future editions can continue to meet your needs.

Glenn Bastian

An Illustrated Review of the DIGESTIVE SYSTEM

1 Essential Nutrients

ESSENTIAL NUTRIENTS / Introduction

THE 6 CLASSES OF NUTRIENTS
An adequate diet includes 6 kinds of nutrients:
(1) Carbohydrates
(2) Lipids
(3) Proteins
(4) Vitamins
(5) Minerals
(6) Water

THE 50 ESSENTIAL NUTRIENTS
Essential nutrients are substances that are required for normal body function, but are not synthesized by the body in adequate amounts. There are approximately 50 substances classified as essential nutrients. Although glucose is an important nutrient, it is not considered an *essential* nutrient because adequate amounts of glucose can be synthesized from other substances; gluconeogenesis is the formation of glucose from noncarbohydrate precursors, such as amino acids, lactic acid, glycerol, and Krebs cycle intermediates.
Water
Amino Acids (9)
Fatty Acids (3)
Vitamins (13)
Minerals (20)
Growth Factors : Choline, Inositol, and Carnitine

SOURCES OF CALORIES
The American Heart Association recommends that an average person who wants to maintain normal blood cholesterol and lipid levels should divide his or her daily intake of calories in the following way: at least 50% from complex carbohydrates; 30% or less from lipids (fats); and 20% or less from proteins. One gram of fat provides 9 calories of energy, while one gram of carbohydrate or protein provides 4 calories of energy. Depending upon weight, gender, and physical acitivity, an average person requires between 2,000 and 3,000 calories per day.

Complex Carbohydrates   50%
Complex carbohydrates (polysaccharides or starches) are large molecules made of hundreds of sugar molecules linked together. Complex carbohydrates are found in many foods including cereals, breads, dry beans, peas, and potatoes.

Lipids (Fats)   30%
The major components of dietary fats are fatty acids. There are three basic types of fatty acids: saturated fats, monounsaturated fats, and polyunsaturated fats. Saturated fats are found mostly in animal fats (butter, lard, dairy products, meats); monounsaturated and polyunsaturated fats are found mostly in vegetables. Another type of lipid, cholesterol, is found in egg yolks, meat, shellfish, whole-milk products, and poultry.

Proteins   20%
Proteins are large molecules made of long chains of amino acids. Foods rich in protein include meats, eggs, milk, and cheese. Proteins found in grains, nuts, legumes, and seeds usually lack one of the essential amino acids.

NUTRITIONAL GUIDELINES

CALORIC VALUE OF BASIC NUTRIENTS

Lipids (Fats) : 9 Calories per gram
Carbohydrates : 4 Calories per gram
Proteins : 4 Calories per gram

THE BASIC DIET

Recommended by the American Heart Association
for the average person who wants to maintain
normal blood cholesterol and lipid levels.

Percent of Total Calories

Complex Carbohydrates 50 %	Lipids (Fats) 30 %	Proteins 20 %

1/3 Saturated Fats
1/3 Monounsaturated Fats
1/3 Polyunsaturated Fats
300 mg of Cholesterol

DIETARY GUIDELINES

Recommended by the Committee on Diet and Health of the
Food and Nutrition Board (under the National Research Council).

(1) Total Fat Intake : reduce to 30 percent or less of total calories.
(2) Complex Carbohydrates : eat 6 or more daily servings.
(3) Fruits and Vegetables : eat 5 or more daily servings.
(4) Proteins : maintain protein consumption at moderate levels.
(5) Exercise : balance food intake with exercise to maintain weight.
(6) Alcohol : do not drink alcohol; or limit the amount to 2 drinks daily.
(7) Salt : limit the amount to 6 grams (slightly less than 1 teaspoon).
(8) Calcium : maintain an adequate calcium intake.
(9) Dietary Supplements : Avoid excess of U.S. RDAs in any one day.
(10) Fluoride : maintain an optimal level of fluoride in the diet.

ESSENTIAL NUTRIENTS / Carbohydrates

Carbohydrates are organic compounds made up of carbon, hydrogen, and oxygen. The ratio of carbon to hydrogen to oxygen is usually 1 : 2 : 1. They are divided into three major groups based on size: monosaccharides, disaccharides, and polysaccharides. Monosaccharides and disaccharides are also called *simple carbohydrates*; polysaccharides are called *complex carbohydrates*.

STRUCTURE
Monosaccharides *(single sugars)*
Glucose, fructose, and galactose.

Disaccharides *(double sugars)*
Maltose (glucose and glucose); sucrose (glucose and fructose); lactose (glucose and galactose).

Polysaccharides
Polysaccharies consist of hundreds or thousands of glucose molecules linked together in long chains. Three important examples of polysaccharides are starch, cellulose, and glycogen.
Starch   Starch is the principal storage form for sugar in plants.
Cellulose   Cellulose is a structural molecule found in plants. Cellulose fibers surround the cell membranes of plant cells, forming a structure called the *cell wall*.
Glycogen   Glycogen is the principal storage form for sugar in higher animals.

SOURCES
Simple Carbohydrates
Maltose is found in malted milk, malted cereals, sprouting grains, and some corn syrups.
Sucrose is found in table sugar, molasses, honey, maple syrup, and fruits.
Lactose is found in milk.

Complex Carbohydrates
Digestible Complex Carbohydrates
Starches are the most abundant carbohydrates in the diet. They are found in many foods including vegetables, cereals, breads, potatoes, and legumes (dry beans and peas).
Indigestible Complex Carbohydrates (Fiber, Bulk, or Roughage)
Cellulose and pectin are the two most important indigestible complex carbohydrates in the diet. Large quantities are found in bran, fruits, whole-grain cereals and breads, and vegetables.

PRODUCTS OF DIGESTION
Monosaccharides   Glucose, fructose, and galactose.

USES
Energy   Carbohydrates are the body's principal energy source (50% of total caloric intake). Catabolism of carbohydrates yields 4 calories per gram; about 200 calories of carbohydrate are stored as glycogen in muscle and liver cells.

Fiber (Bulk or Roughage)
The following are uses of fiber: regularity (softens stool, solidifies watery stool, increases bulk, decreases transit time, prevents diverticula); satiety (sense of fullness); blood glucose modulation; lowers serum lipid (binds with fiber); slows carbohydrate absorption.

NUTRIENT REQUIREMENTS
Complex Carbohydrates   At least 50% of daily caloric intake.
Fiber   15 - 20 grams daily.

CARBOHYDRATES

Monosaccharides

Glucose $C_6H_{12}O_6$

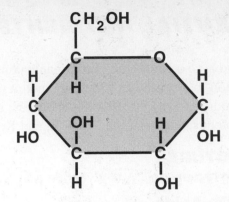

Abbreviated Formulas

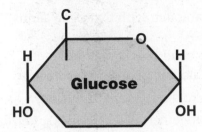 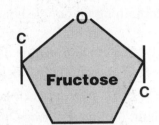 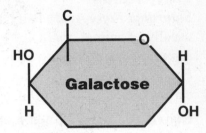

Disaccharides

Maltose   **Sucrose**

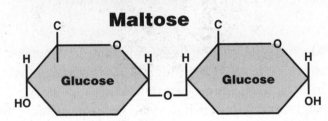 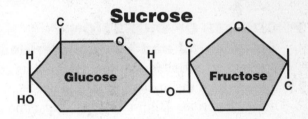

Polysaccharides

Starch

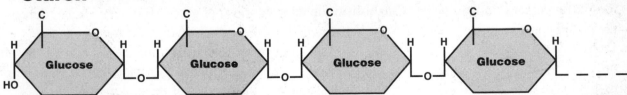

Cellulose

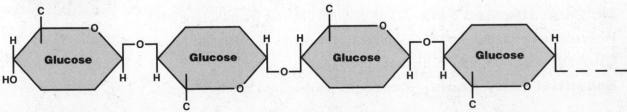

5

ESSENTIAL NUTRIENTS / Lipids

Lipids (also called fats) are organic compounds that are insoluble in water; they are made up of carbon, hydrogen, and a small amount of oxygen. Lipids are a diverse group of substances including triglycerides, phospholipids, steroids (cholesterol, cortisol, sex hormones), eicosanoids, carotenes, and vitamins A, D, E, and K. The major lipid in foods is triglyceride (neutral fat).

STRUCTURE

Triglyceride A triglyceride is a molecule of glycerol with three fatty acids attached.

Fatty Acids

Saturated Fatty Acid Fatty acid containing no double bonds (saturated with hydrogen atoms); usually of animal origin.

Monounsaturated Fatty Acid Fatty acid containing one double bond.

Polyunsaturated Fatty Acid Fatty acid containing two or more double bonds between its carbon atoms; usually of plant origin; most vegetable oils.

Steroid A class of lipids containing four carbon rings linked together; *cholesterol* is a steroid.

SOURCES

Visible Fats Butter, margarine, oil, cream, and marbled red meats.

Hidden Fats Nuts, coconuts, olives, chocolate, cheeses, and processed meats.

Cholesterol Egg yolk, shellfish, organ meats (beef liver), and whole milk products.

PRODUCTS OF DIGESTION

Fatty Acids and Monoglycerides Mechanical agitation in the small intestine breaks up large fat droplets into smaller droplets coated with bile salts, exposing more surface area for the action of the enzyme *lipase*. Lipase splits triglyceride, forming two fatty acids and one monoglyceride. Bile salts, fatty acids, and monoglycerides cluster in water-soluble particles called *micelles* for transport in the intestine. Fatty acids and monoglycerides released from micelles diffuse into epithelial cells lining the digestive tract. Triglycerides are resynthesized inside epithelial cells and coated with proteins, forming complexes called *chylomicrons*. Chylomicrons are secreted from epithelial cells and enter intestinal lacteals (lymphatic capillaries).

USES

Energy Lipids are the major fuel reserve of the body; 100,000 calories are stored as triglycerides in fat cells (adipose tissue). Catabolism yields 9 calories per gram. Between meals (during the postabsorptive stage) most cells use fatty acids for the production of ATP (energy).

Other Uses Steroid hormones and bile salts are synthesized from cholesterol; adipose tissue provides insulation and cushioning; oils secreted by sebaceous glands lubricate the skin and help to condition the hair; phospholipids and cholesterol are constituents of plasma membranes.

NUTRIENT REQUIREMENTS *50 to 100 grams daily (30% of total caloric intake)*

Saturated Fats 1/3 of the total calories from dietary fats.

Monounsaturated Fats 1/3 of the total calories from dietary fats.

Polyunsaturated Fats 1/3 of the total calories from dietary fats.

Cholesterol 300 mg (milligrams) per day.

Essential Fatty Acids Arachidonic, linoleic, and linolenic acids.

LIPIDS

Triglyceride (example : Tristearin)

Saturated Fatty Acid Stearic Acid $C_{17}H_{35}COOH$

Monounsaturated Fatty Acid Oleic Acid $C_{17}H_{33}COOH$

Polyunsaturated Fatty Acid Trilinolein $C_{17}H_{31}COOH$

Cholesterol

Glycerol   Monoglyceride

7

ESSENTIAL NUTRIENTS / Proteins

Proteins are very large, complex organic compounds made up of carbon, hydrogen, nitrogen, and sometimes sulfur. They consist of chains of amino acids.

STRUCTURE

Amino Acids There are about 20 different amino acids. Each amino acid consists of a central carbon atom to which 4 groups are attached : a hydrogen atom, an amino group ($-NH_2$), a carboxyl group ($-COOH$), and a variable group designated R (gives uniqueness to amino acid).
Essential Amino Acids Essential amino acids must be included in the daily diet. They include isoleucine, leucine, lysine, methionine, phenylalanine, threonine, tryptophan, valine; (arginine and histidine cannot be synthesized in amounts that are adequate for growing children).
Nonessential Amino Acids Nonessential amino acids are amino acids that can be synthesized in the body. They include alanine, asparagine, aspartic acid, cysteine, glutamic acid, glutamine, glycine, proline, serine, and tyrosine.

Dipeptide Two amino acids linked by a covalent bond (called a peptide bond).

Peptide A chain of 50 or fewer amino acids linked by peptide bonds.

Protein A chain of over 50 amino acids linked by peptide bonds. The 20 different amino acids arranged in different sequences form roughly 100,000 different kinds of proteins; their complex shapes are determined primarily by interactions of the variable groups along the chain.

SOURCES

Complete Proteins Complete proteins contain all of the essential amino acids; they are usually of animal origin (meat, eggs, milk, cheese).

Incomplete Proteins Incomplete proteins lack one or more of the essential amino acids; they are usually of plant origin (grains, nuts, legumes, seeds).

Complementary Proteins Complementary proteins are combinations of plant proteins that include all of the essential amino acids. Examples: peanut butter and bread; rice and beans.

PRODUCTS OF DIGESTION

Amino Acids Essential and nonessential amino acids.

USES

Structural Roles Proteins are the major structural material in most tissues. *Integral* and *peripheral proteins* are important constituents of plasma membranes; *collagen* is a constituent of bones and connective tissues; the protein *keratin* is found in the skin, hair, and fingernails.
Functional Roles Antibodies (immunoglobulins); complement proteins; enzymes; hormones; buffers; storage proteins (ferritin, hemosiderin, and myoglobin); transport proteins (hemoglobin, transferrin, and lipoproteins); contractile proteins (actin and myosin); membrane transport proteins; membrane pores; surface antigens; clotting factors; and albumin.

NUTRIENT REQUIREMENTS

U.S. RDA : 0.8 gram daily per kilogram of body weight. (1 kilogram = 1,000 grams = 2.2 lbs.)
Average Male (70 kg or 154 lb) : 56 grams daily.
Average Female (55 kg or 120 lb) : 44 grams daily.

PROTEINS

Amino Acids
20 different amino acids are found in proteins; 8 amino acids are essential (must be in dietary foods).

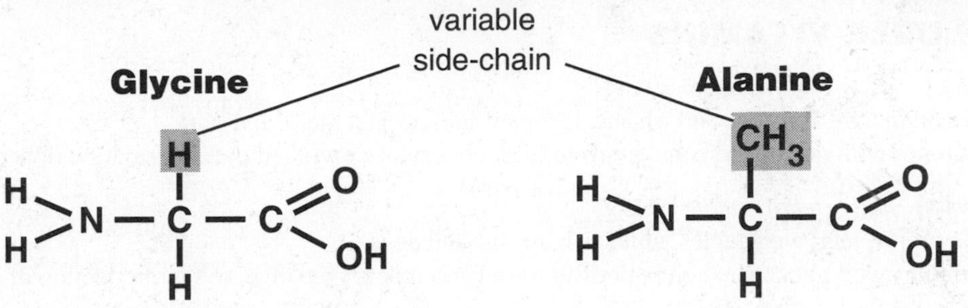

variable side-chain

Glycine

Alanine

H—N—C—C=O
    | | OH
    H   H
    |
    H

(H highlighted as side-chain)

H—N—C—C=O
    | | OH
    H   H

(CH_3 highlighted as side-chain)

Dipeptide
2 amino acids linked by a peptide bond.

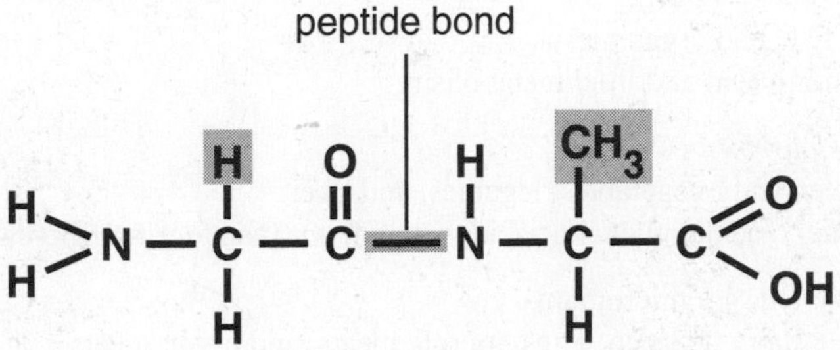

peptide bond

H—N—C—C=O ... N—C—C=O
    | | | | OH
    H   H   H   H

(H highlighted, CH_3 highlighted)

Peptide
A peptide consists of fewer than 50 amino acids.

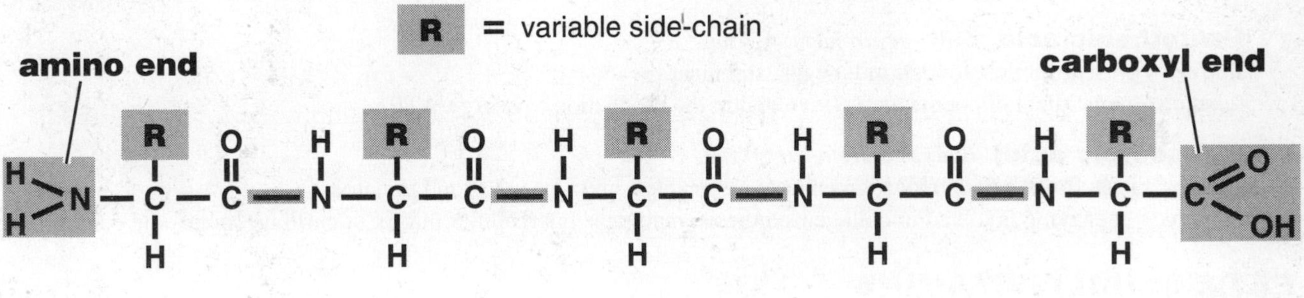

R = variable side-chain

amino end

carboxyl end

H—N—C—C=O — N—C—C=O — N—C—C=O — N—C—C=O — N—C—C=O
    | | | | | | | | | | OH
    H   H   H   H   H   H   H   H   H   H

(R highlighted at each residue)

Protein
A protein consists of more than 50 amino acids.

● = amino acid

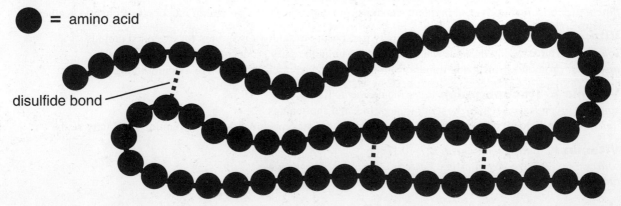

disulfide bond

9

ESSENTIAL NUTRIENTS / Vitamins

Vitamins are small organic substances that are required in minute amounts (milligrams or micrograms per day) to maintain growth and metabolism; most vitamins cannot be synthesized by the body. Most of the water-soluble vitamins are components of coenzymes; the fat-soluble vitamins have a variety of functions.

WATER–SOLUBLE VITAMINS

B₁ (thiamine) *daily requirement:* 1.9 mg.
sources: whole grains (husk of rice and wheat), legumes, milk, organ meat, and pork.
functions: combines with phosphate compounds to form coenzymes involved in the formation of acetyl CoA.

B₂ (riboflavin) *daily requirement:* 1.5 mg.
sources: whole grains, leafy vegetables, milk, fish, meats, and eggs.
functions: combines with phosphate compounds to form FAD (coenzyme involved in production of energy/ATP).

Niacin (nicotinic acid) *daily requirement:* 14.6 mg.
sources: whole grains, potatoes, legumes, lean meats, and liver.
functions: constituent of NAD and NADP (coenzymes involved in production of energy/ATP).

B₆ (pyridoxine) *daily requirement:* 1.42 mg.
sources: whole grains, vegetables, fish, and organ meats.
functions: coenzyme involved in amino acid and lipid metabolism.

Folacin (folic acid) *daily requirement:* 0.1 mg.
sources: whole-wheat products, green leafy vegetables, legumes, and liver.
functions: coenzyme involved in amino acid and nucleic acid metabolism (transfer of single-carbon units).

B₁₂ (cobalamin) *daily requirement:* 4.5 micrograms.
sources: (not present in plant foods), dairy products, eggs, muscle meats, and organ meats.
functions: coenzyme involved in nucleic acid metabolism (transfer of single-carbon units).

Biotin *daily requirement:* 0.1 - 0.2 mg.
sources: (synthesis by intestinal bacteria), vegetables, legumes, egg yolk, and meats.
functions: coenzyme involved in fat synthesis, amino acid metabolism, and glycogen formation.

Pantothenic acid *daily requirement:* 4.7 mg.
sources: (present in most foods), milk, eggs, and meat.
functions: constituent of coenzyme A (involved in the production of energy/ATP).

C (ascorbic acid) *daily requirement:* 60 mg.
sources: citrus fruits, salad greens, green leafy vegetables, green peppers, and tomatoes.
functions: coenzyme involved in collagen synthesis; maintains intercellular matrix of cartilage, bone, and dentin.

FAT–SOLUBLE VITAMINS

Vitamin A *daily requirement:* 1 mg.
sources: **Beta-carotene (Provitamin A)** is widely distributed in dark green leafy and yellow vegetables.
        Retinal is present in milk, cheese, fortified margarine, and butter.
functions: constituent of visual pigment (rhodopsin); maintenance of epithelium; mucopolysaccharide synthesis.

Vitamin D *daily requirement:* needed only if exposure to sunlight is limited for extended periods.
sources: (synthesized in the skin when exposed to sunlight), dairy products, eggs, fish oils, and liver.
functions: promotes growth and mineralization of bones; increases absorption of calcium in digestive tract.

Vitamin E (tocopherols) *daily requirement:* 12 mg.
sources: green leafy vegetables, seeds, margarine, and shortenings.
functions: antioxidant to prevent cell membrane damage; prevents breakdown of vitamin A and fatty acids.

Vitamin K *daily requirement:* 0.7 - 0.14 mg.
sources: (synthesized by intestinal bacteria), green leafy vegetables.
functions: essential for liver synthesis of prothrombin and other clotting factors.

VITAMINS

Water-soluble Vitamins

Water-soluble vitamins include the B vitamins and vitamin C.

Vitamin C

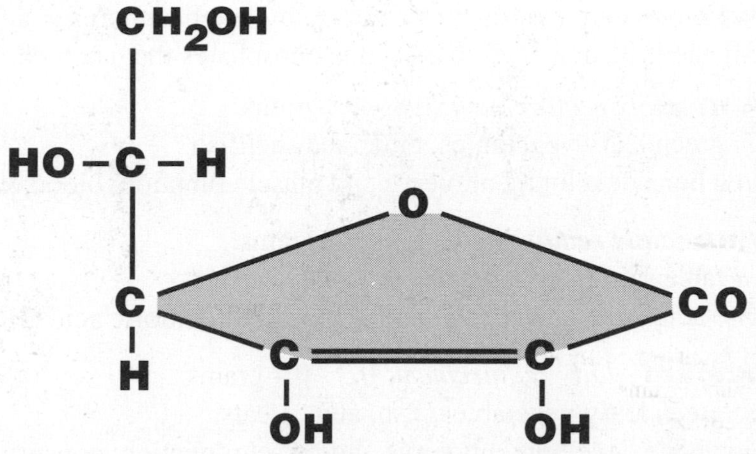

$$CH_2OH$$
$$HO-C-H$$

(structure showing ring with O, C=C, OH, OH, CO, and C-H)

Fat-soluble Vitamins

Fat-soluble vitamins include vitamins A, D, E, and K.

Vitamin D

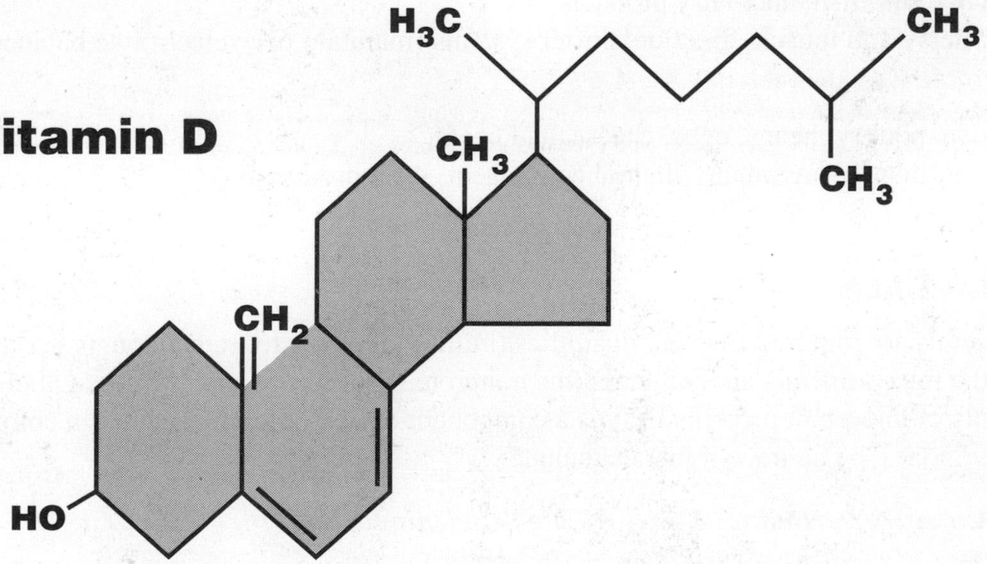

(steroid-like ring structure with H_3C, CH_3, CH_3, CH_3, CH_2, CH_3, and HO groups)

ESSENTIAL NUTRIENTS / Minerals

Minerals are inorganic substances with functions that are essential to life.

BULK MINERALS

Bulk minerals are required in relatively large amounts (in the neighborhood of grams). Often they have more than one function. For example, phosphorus is an important constituent of bones and teeth and is also needed to make the phosphates that are a component of ATP.

Calcium *daily requirement:* 0.8 - 1.2 grams.
sources: green leafy vegetables, milk, and shellfish.
functions: bone development; nerve and muscle function; blood clotting.

Chlorine *daily requirement:* 1.7 - 5.1 grams.
sources: table salt, fish, and dairy products.
functions: acid-base balance; water balance; hydrochloric acid (HCl) formation.

Magnesium *daily requirement:* 0.3 - 0.4 grams.
sources: green leafy vegetables, fish, and cereals.
functions: bone development; nerve and muscle function; constituent of coenzymes.

Phosphorus *daily requirement:* 0.8 - 1.2 grams.
sources: fish, poultry, meats, and dairy products.
functions: bone development,; nerve and muscle function; buffer systems (maintain pH); enzyme component; energy transfer (ATP production).

Potassium *daily requirement:* 1.9 - 5.6 grams.
sources: fruits, nuts, meats, and dairy products.
functions: nerve and muscle function.

Sodium *daily requirement:* 1.1 - 3.3 grams.
sources: table salt, fish, and dairy products.
functions: nerve and muscle function; buffer systems (maintain pH); electrolyte balance.

Sulfur
sources: fish, poultry, beans, eggs, cheese, and beef.
functions: hormone component; vitamin component; ATP production.

TRACE MINERALS

Trace minerals are required in small quantities (milligram or microgram amounts per day). Many of the trace minerals are cofactors (the nonprotein portion of an enzyme). Others are components of important proteins (iron is a component of hemoglobin; iodine is a component of thyroid hormones). The trace minerals include:

Copper Cu	*Iodine I*	*Selenium Se*	*Zinc Zn*
Cobalt Co	*Iron Fe*	*Silicon Si*	
Chromium Cr	*Manganese Mn*	*Tin Sn*	
Fluorine F	*Molybdenum Mo*	*Vanadium V*	

MINERALS

Bulk Minerals

Bulk minerals are required in amounts in excess of 100 mg/day.

Name and Symbol	Sources	Functions
Calcium Ca	green leafy vegetables, milk, and shellfish	bone development, nerve and muscle function, blood clotting
Chlorine Cl	table salt, fish, and dairy products	acid-base balance, water balance, HCl formation
Magnesium Mg	green leafy vegetables, fish, and cereals	bone development, nerve and muscle function, constituent of coenzymes
Phosphorus P	fish, poultry, meats, and dairy products	bone development, nerve and muscle function, buffer systems, enzyme component, energy transfer
Potassium K	Fruits, nuts, meats, and dairy products	nerve and muscle function
Sodium Na	table salt, fish, and dairy products	nerve and muscle function, buffer system, electrolyte balance
Sulfur S	fish, poultry, beans, eggs, cheese, and beef	hormone component, vitamin component, ATP production

Trace Minerals

Trace minerals are required in relatively small amounts: less than 100 mg/day.

Copper Cu	Iodine I	Selenium Se	Zinc Zn
Cobalt Co	Iron Fe	Silicon Si	
Chromium Cr	Manganese Mn	Tin Sn	
Fluorine F	Molybdenum Mo	Vanadium V	

2 Structures and Functions

STRUCTURES AND FUNCTIONS / Overview

STRUCTURAL ORGANIZATION

The organs of the digestive system are divided into two main groups: the gastrointestinal tract and the accessory structures.

Gastrointestinal Tract (GI Tract)

The gastrointestinal tract is also called the GI tract or alimentary canal. It is a continuous tube running through the ventral body cavity from the mouth to the anus. The major organs of the GI tract are the following :

(1) oral cavity (mouth)   *(4) stomach*
(2) pharynx (throat)   *(5) small intestine*
(3) esophagus   *(6) large intestine*

Accessory Structures

The major accessory structures of the digestive system are the following:

(1) teeth   *(4) liver*
(2) tongue   *(5) gallbladder*
(3) salivary glands   *(6) pancreas*

DIGESTIVE PROCESSES

The digestive system carries out five basic processes.

(1) Ingestion (Eating) Taking food into the mouth.

(2) Movement (Motility) The movement of food through the GI tract by rhythmic contractions.

(3) Digestion The breaking down of food into molecules that are small enough to be absorbed.

(4) Absorption The passage of digested food from the GI tract into the blood and lymph.

(5) Defecation The elimination of indigestible substances (feces) from the GI tract.

REGULATION OF DIGESTIVE PROCESSES

Neural Regulation

Enteric Nervous System The GI tract has its own network of nerves called the enteric nervous system, which consists of autonomic nerve fibers. The *myenteric plexus* is located in a layer of the GI tract wall called the muscularis; the *submucosal plexus* is located in another layer of the GI tract called the submucosa. These nerve plexuses control GI tract motility.

Hormonal Regulation

Four major digestive hormones help control the digestive processes. These hormones are secreted by *enteroendocrine cells* that are scattered among the epithelial cells lining the GI tract.

(1) Gastrin Stimulates secretion of gastric juice and gastric motility.

(2) Secretin Stimulates secretion of bicarbonate ions by the pancreas and liver.

(3) Cholecystokinin (CCK) Stimulates secretion of digestive enzymes by the pancreas and the ejection of bile from the gallbladder.

(4) Gastric Inhibitory Peptide (GIP) Inhibits secretion of gastric juice and gastric emptying.

GASTROINTESTINAL TRACT

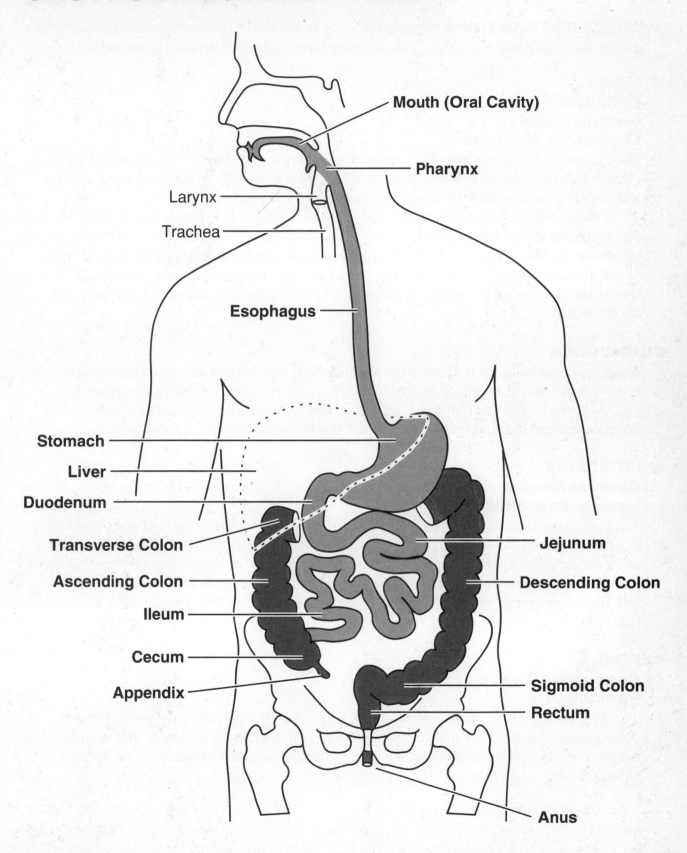

Mouth (Oral Cavity)

Pharynx

Larynx

Trachea

Esophagus

Stomach

Liver

Duodenum

Transverse Colon

Ascending Colon

Ileum

Cecum

Appendix

Jejunum

Descending Colon

Sigmoid Colon

Rectum

Anus

17

The wall of the GI tract from the esophagus to the end of the large intestine has the same basic arrangement of four tissue layers. The following layers are listed from the inside out.

MUCOSA
Epithelium
Lamina Propria
Muscularis Mucosae

The mucosa is the mucous membrane that lines the GI tract. It consists of epithelium (in direct contact with the contents of the GI tract), areolar connective tissue (called the lamina propria), and smooth muscle (called the muscularis mucosae).

In the mouth, esophagus, and anal canal the epithelial layer is *stratified squamous epithelium*; throughout the rest of the GI tract it is *simple columnar epithelium*. The *lamina propria* is made of areolar connective tissue containing many blood and lymphatic vessels; it also contains lymphatic nodules. The *muscularis mucosae* contains smooth muscle fibers that pull the mucous membrane of the intestine into small folds, which increase the surface area for digestion and absorption.

SUBMUCOSA

The submucosa binds the mucosa to the muscularis. It consists of areolar connective tissue and contains many blood vessels, a network of autonomic nerves (the *submucosal plexus* or the *plexus of Meissner*), and *submucosal glands*. It also contains *lymphatic nodules* composed of lymphocytes and macrophages (cells involved in defense mechanisms against microbes).

MUSCULARIS
Circular Muscle
Longitudinal Muscle

The muscularis consists mostly of muscle fibers (smooth or skeletal muscle fibers, depending upon the region of the GI tract). It contains a network of autonomic nerve fibers (the *myenteric plexus* or *plexus of Auerbach*). The myenteric plexus consists of both sympathetic and parasympathetic nerve fibers. It is located between the two muscle sublayers and controls GI tract motility. Blood and lymph vessels are present in the connective tissue between the two muscle sublayers.

SEROSA
Connective Tissue
Epithelium

The serosa is the outermost layer of most portions of the GI tract. It is a serous membrane composed of a layer of *simple squamous epithelium* (mesothelium) and underlying *areolar connective tissue*. In the portions of the GI tract that are below the diaphragm, the serosa is called the *visceral peritoneum*.

LAYERS OF THE GI TRACT

The wall of the GI tract from the esophagus to the anus has the same basic arrangement of tissues.

The four layers of the GI tract from the inside out are :

(1) Mucosa   (3) Muscularis

(2) Submucosa (4) Serosa

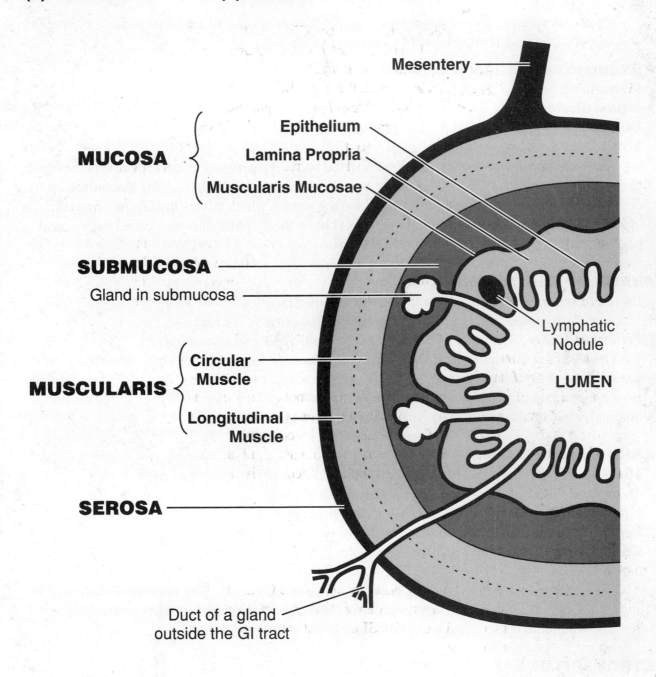

Mesentery

MUCOSA
- Epithelium
- Lamina Propria
- Muscularis Mucosae

SUBMUCOSA
Gland in submucosa

Lymphatic Nodule

LUMEN

MUSCULARIS
- Circular Muscle
- Longitudinal Muscle

SEROSA

Duct of a gland outside the GI tract

STRUCTURES AND FUNCTIONS / Peritoneum

The peritoneum is the largest serous membrane of the body. It lines the abdominal cavity and covers some of the viscera (abdominal organs).

SEROUS MEMBRANES
Histology
Simple Squamous Epithelium (called Mesothelium)
Loose Connective Tissue
    All serous membranes consist of two layers of tissues: a layer of simple squamous epithelium and an underlying supporting layer of connective tissue.

Examples of Serous Membranes
Pericardium
Pleural Membranes
Peritoneum
Serosa
    There are four major serous membranes in the body: the pericardium, which covers the heart; the pleural membrane, which covers the lungs; the peritoneum, which lines the abdominal cavity and covers the viscera (abdominal organs); and the serosa, which covers the GI tract from the esophagus to the anus (the portion of the serosa below the diaphragm is also called the visceral peritoneum).

LAYERS OF PERITONEUM
Parietal Peritoneum   Lines the wall of the abdominal cavity.
Visceral Peritoneum   Covers some of the abdominal organs (viscera).
Peritoneal Cavity   The potential space between the parietal and visceral layers.
    The peritoneum is a thin, transparent serous membrane located in the abdominal cavity. It consists of two layers separated by a potential space called the peritoneal cavity. The peritoneal cavity contains serous fluid, which lubricates the surfaces of the membranes, enabling the abdominal organs (viscera) to move with a minimum of friction.

PERITONEAL FOLDS
Mesentery   Binds the small intestine to the abdominal wall.
Mesocolon   Binds the large intestine to the abdominal wall.
Falciform Ligament   Attaches the liver to the diaphragm and the anterior abdominal wall.
Lesser Omentum   Suspends the stomach and duodenum from the liver.
Greater Omentum   Forms an apronlike structure, draping over the transverse colon and small intestine. It is the largest peritoneal fold.
    The peritoneum has large folds that weave between the viscera. The peritoneal folds bind the abdominal organs to each other and to the walls of the abdominal cavity. They contain blood vessels, lymphatic vessels, and nerves that supply the abdominal organs.

RETROPERITONEAL ORGANS
The kidneys and pancreas lie against the posterior abdominal wall and are covered by peritoneum on their anterior surfaces only. They are behind the peritoneum (*retro* = backward or located behind). Some portions of the large intestine are retroperitoneal.

PERITONEUM

Midsagittal Section

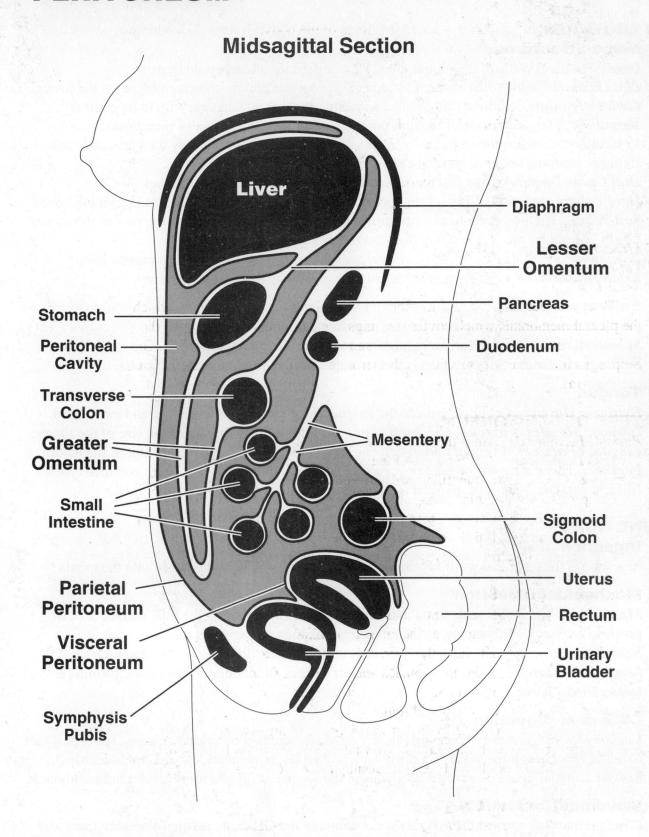

Liver

Diaphragm

Lesser Omentum

Pancreas

Stomach

Duodenum

Peritoneal Cavity

Transverse Colon

Greater Omentum

Mesentery

Small Intestine

Sigmoid Colon

Parietal Peritoneum

Uterus

Rectum

Visceral Peritoneum

Urinary Bladder

Symphysis Pubis

STRUCTURES AND FUNCTIONS / Mouth (Oral Cavity)

STRUCTURES

Basic Structures

Cheeks   Lateral walls of the mouth; lined by stratified squamous epithelium.

Lips (Labia)   Fleshy folds covered by skin on the outside and mucous membrane on the inside.

Labial Frenulum   Midline fold of mucous membrane that attaches each lip to its gum.

Vermilion   Transition zone of the lips, where the skin meets the mucous membrane.

Vestibule   Space bounded externally by the cheeks and lips, internally by the gums and teeth.

Fauces   Opening between the oral cavity and the pharynx (throat).

Oral Cavity Proper   Space that extends from the gums and teeth to the fauces.

Hard Palate   Anterior portion of the roof of the mouth; formed by maxillae and palatine bones.

Soft Palate   Arch-shaped muscular partition, forming posterior portion of the roof of the mouth.

Uvula   Muscular process hanging from the free border of the soft palate.

Palatopharyngeal Arch   The palatopharyngeus muscle covered by mucous membrane.

Palatoglossal Arch   The palatoglossus muscle covered by mucous membrane.

Salivary Glands

Parotid Glands   Salivary glands located anterior and inferior to the ears.

Submandibular Glands   Salivary glands located beneath the base of the tongue.

Sublingual Glands   Salivary glands located superior to the submandibular glands.

Tongue

Extrinsic Muscles   Originate outside the tongue and insert into it; maneuver and move food.

Intrinsic Muscles   Originate and insert within the tongue; alter the shape and size of the tongue.

Lingual Frenulum   Midline fold of mucous membrane; attaches tongue to floor of the mouth.

Papillae   Tiny elevations on the upper surface and sides of the tongue. There are three types: filiform papillae, fungiform papillae, and circumvallate papillae (contain taste buds).

FUNCTIONS

Ingestion (Eating.)

Ingestion is the technical word for eating, the process by which food is taken into the mouth.

Mechanical Digestion

Mastication   Chewing breaks food into smaller particles, increasing the total surface area exposed to enzymes. This increases the efficiency of chemical digestion.

Saliva   Saliva secreted by the salivary glands moistens and lubricates the food particles.

Mixing or Churning   Movements of the tongue help to mix the food with saliva, forming a food mass called a *bolus*.

Chemical Digestion

Carbohydrates   The enzyme called salivary amylase, secreted by salivary glands, splits polysaccharides (starches) into smaller fragments (disaccharides, trisaccharides, and alpha-dextrins).

Lipids   Lingual lipase is secreted by glands in the tongue; it digests triglycerides in the stomach.

Movement or Motility

(Smooth muscle contractions that mix the contents of the GI tract with digestive secretions and move the partially digested food through the tract from the mouth to the anus.)

Deglutition or Swallowing   The swallowing reflex is initiated when the bolus is forced to the back of the oral cavity by movement of the tongue upward and backward against the palate.

MOUTH (Oral Cavity)

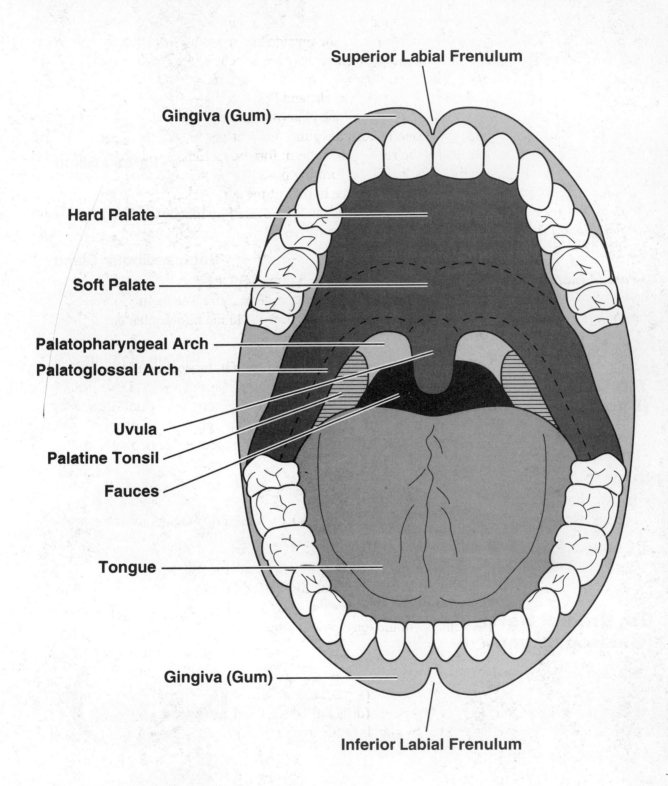

Superior Labial Frenulum

Gingiva (Gum)

Hard Palate

Soft Palate

Palatopharyngeal Arch

Palatoglossal Arch

Uvula

Palatine Tonsil

Fauces

Tongue

Gingiva (Gum)

Inferior Labial Frenulum

SALIVARY GLANDS

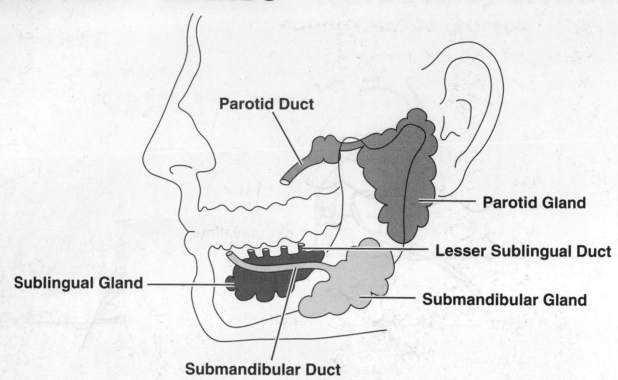

Parotid Duct

Parotid Gland

Lesser Sublingual Duct

Sublingual Gland

Submandibular Gland

Submandibular Duct

7th Cranial Nerve (Facial)

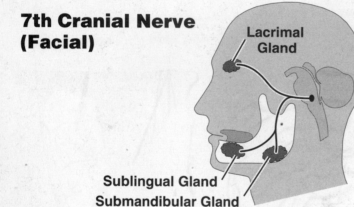

Lacrimal Gland

Sublingual Gland
Submandibular Gland

Tongue (anterior 2/3)

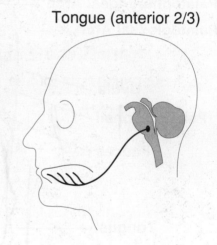

9th Cranial Nerve (Glossopharyngeal)

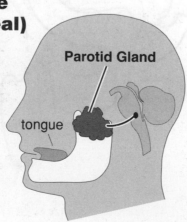

Parotid Gland

tongue

Tongue (posterior 1/3)
Proprioceptors (swallowing)
Pressure Receptors (carotid sinus)

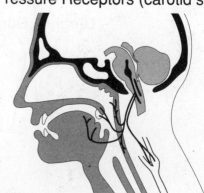

TONGUE

Dorsum of the Tongue

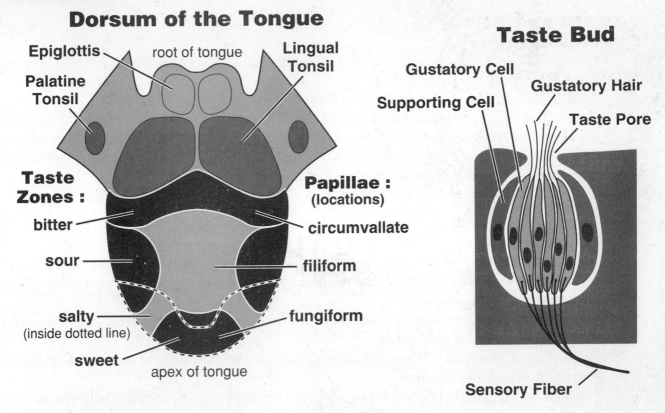

Epiglottis

root of tongue

Lingual Tonsil

Palatine Tonsil

Taste Zones :

bitter

sour

salty
(inside dotted line)

sweet

apex of tongue

Papillae :
(locations)

circumvallate

filiform

fungiform

Taste Bud

Gustatory Cell

Supporting Cell

Gustatory Hair

Taste Pore

Sensory Fiber

Circumvallate Papilla

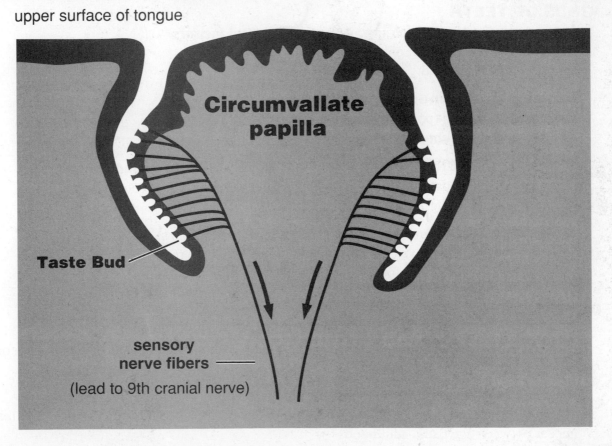

upper surface of tongue

Circumvallate papilla

Taste Bud

sensory nerve fibers

(lead to 9th cranial nerve)

TOOTH ANATOMY
Principal Parts
Crown The exposed portion above the level of the gums. It contains the *pulp cavity* (a cavity filled with pulp).

Neck The junction line between the crown and the roots.

Roots The portion embedded in the alveolar bone of the jaw. Narrow extensions of the pulp cavity that extend through the root are called *root canals*. Each root canal has an opening at its base called the *apical foramen* through which blood vessels, lymphatic vessels, and nerves enter and exit the tooth. A tooth may have one to three roots.

Composition
Dentin Teeth are composed primarily of dentin, a calcified connective tissue. It gives a tooth its shape and rigidity.

Enamel Enamel is a hardened substance consisting primarily of calcium phosphate and calcium carbonate. It covers the dentin of the crown and protects the tooth from the wear of chewing and the corrosive effects of acid.

Pulp Pulp is a connective tissue containing blood vessels, lymphatic vessels, and nerves. It fills the pulp cavity and the root canals.

Cementum Cementum is a bonelike substance that covers the dentin of the root. It attaches the root to the periodontal ligament.

Periodontal Ligament Periodontal ligament is a dense fibrous connective tissue that attaches the cementum to the walls of the tooth socket. It holds the teeth in position and absorbs shock.

KINDS OF TEETH
Incisors (2 pairs in each jaw; central and lateral incisors)
structure : chisel-shaped; one root.

function : cut into food.

Cuspids or Canines (1 pair in each jaw)
structure : one cusp (pointed surface) and one root.

function : tear and shred food.

Premolars or Bicuspids (2 pairs in each jaw; first and second premolars)
structure : two cusps and one root (upper first premolars have two roots).

function : crush and grind food.

Molars (3 pairs in each jaw; first, second, and third molars)
structure : flattened crowns with prominent ridges and three roots.

function : crush and grind food.

DENTITIONS (SETS)
Deciduous Teeth (Milk Teeth or Baby Teeth)   20 teeth
The first deciduous teeth begin to erupt when a baby is about 6 months old. Each month one pair appears until 20 teeth are present.

Permanent Teeth   32 teeth
Deciduous teeth are lost when a child is between 6 and 12 years old. The 32 permanent teeth appear between age 6 and adulthood.

TOOTH ANATOMY

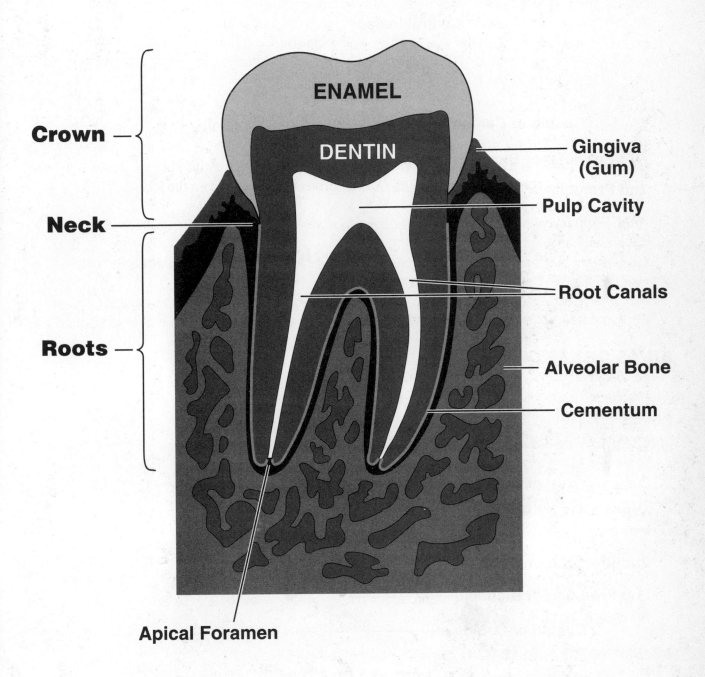

Crown

Neck

Roots

ENAMEL

DENTIN

Gingiva (Gum)

Pulp Cavity

Root Canals

Alveolar Bone

Cementum

Apical Foramen

PERMANENT TEETH (32 Teeth)
8 Incisors, 4 Cuspids, 8 Premolars, and 12 Molars

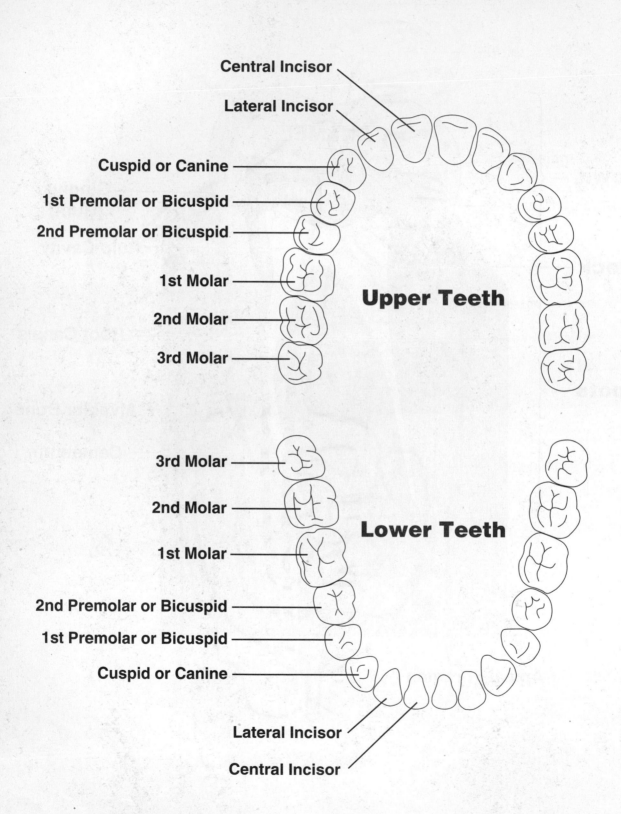

Central Incisor

Lateral Incisor

Cuspid or Canine

1st Premolar or Bicuspid

2nd Premolar or Bicuspid

1st Molar

2nd Molar

3rd Molar

Upper Teeth

3rd Molar

2nd Molar

1st Molar

Lower Teeth

2nd Premolar or Bicuspid

1st Premolar or Bicuspid

Cuspid or Canine

Lateral Incisor

Central Incisor

28

PREMOLAR WITH BLOOD VESSELS

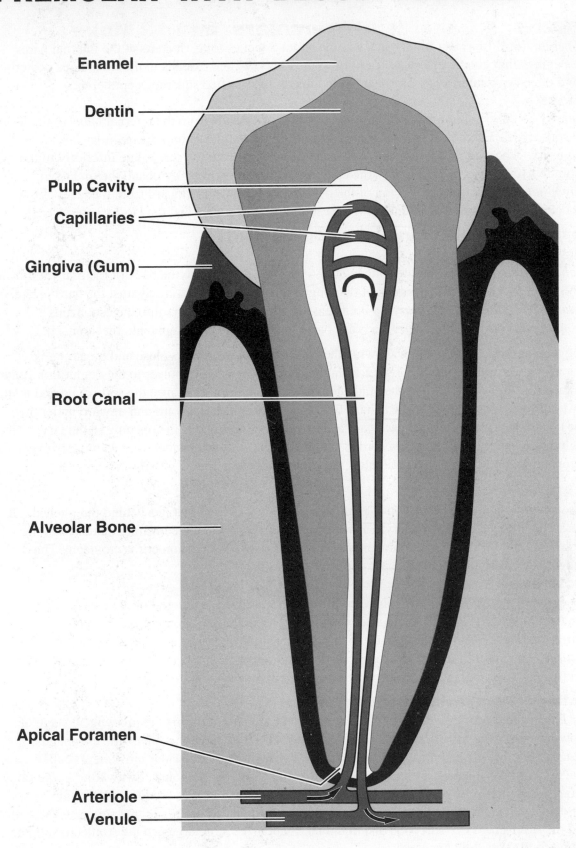

Enamel

Dentin

Pulp Cavity

Capillaries

Gingiva (Gum)

Root Canal

Alveolar Bone

Apical Foramen

Arteriole

Venule

PHARYNX

The pharynx (or throat) is a funnel-shaped tube about 5 inches long. It starts at the internal nares (two openings that connect the nasal cavities to the pharynx) and extends to the esophagus posteriorly and the larynx anteriorly. The pharynx is lined with stratified squamous epithelium.

Portions

Nasopharynx   The uppermost portion of the pharynx; it extends from the internal nares to the level of the soft palate. Each of the lateral walls has an opening that leads into the *auditory tubes*.

Oropharynx   The middle portion of the pharynx; it extends from the soft palate inferiorly to the level of the hyoid bone. The *fauces* is the opening between the oropharynx and the oral cavity.

Laryngopharynx   The lowest portion of the pharynx; it extends downward from the level of the hyoid bone and becomes continuous with the esophagus posteriorly and the larynx anteriorly.

Digestive Function : Swallowing (Deglutition)

Swallowing is divided into three stages :

(1) Voluntary Stage: a bolus (soft, rounded mass of food) is forced to the back of the oral cavity and into the oropharynx by the movement of the tongue upward and backward against the hard palate.

(2) Pharyngeal Stage (involuntary): a bolus passes through the pharynx into the esophagus.

(3) Esophageal Stage (involuntary): a bolus passes through the esophagus into the stomach.

During the pharyngeal stage of swallowing the respiratory passageways close and breathing is temporarily interrupted. Sensory receptors in the oropharynx send impulses to the deglutition center in the medulla and lower pons of the brain stem. Returning impulses cause the soft palate and uvula to move upward, closing off the nasopharynx; the larynx is pulled forward and upward under the tongue, and the epiglottis moves backward and downward, sealing off the opening in the larynx (the rima glottidis) and pulling the vocal cords together. This action also widens the opening between the laryngopharynx and esophagus, facilitating the movement of the bolus into the esophagus.

ESOPHAGUS

The esophagus is a muscular, collapsible tube about 10 inches long that lies behind the trachea. It begins at the inferior end of the laryngopharynx, passes through the mediastinum anterior to the vertebral column, pierces the diaphragm, and ends in the superior portion of the stomach. The esophagus secretes mucus and transports food to the stomach.

Structures

Upper Esophageal Sphincter   A sphincter muscle (valve) at the upper end of the esophagus.

Lower Esophageal Spincter   A sphincter muscle (valve) at the lower end of the esophagus.

Esophageal Hiatus   The opening in the diaphragm through which the esophagus passes.

Adventitia   The outer layer of the esophagus is called the adventitia (rather than the serosa). The connective tissue of this layer is not covered by epithelium (mesothelium).

Function : Swallowing (Deglutition)

Upper Esophageal Sphincter Opens   The elevation of the larynx during the pharyngeal stage of swallowing causes the upper esophageal sphincter to relax, and the bolus enters the esophagus.

Peristalsis   Food is pushed through the esophagus by involuntary muscular movements called peristalsis. Circular muscles above the bolus contract, squeezing the bolus downward. At the same time longitudinal muscles just below the bolus contract, shortening this section and pushing its walls outward so it can receive the bolus. The contractions are repeated in a wave that squeezes the bolus toward the stomach. Very soft food and liquid takes about 1 second to reach the stomach; solid or semisolid food takes 4 to 8 seconds.

Lower Esophageal Sphincter Opens   During swallowing the lower esophageal sphincter relaxes, allowing the bolus to pass from the esophagus into the stomach.

SWALLOWING (DEGLUTITION)

Stage 1
Voluntary Stage

Stages 2 and 3
Pharyngeal Stage and Esophageal Stage

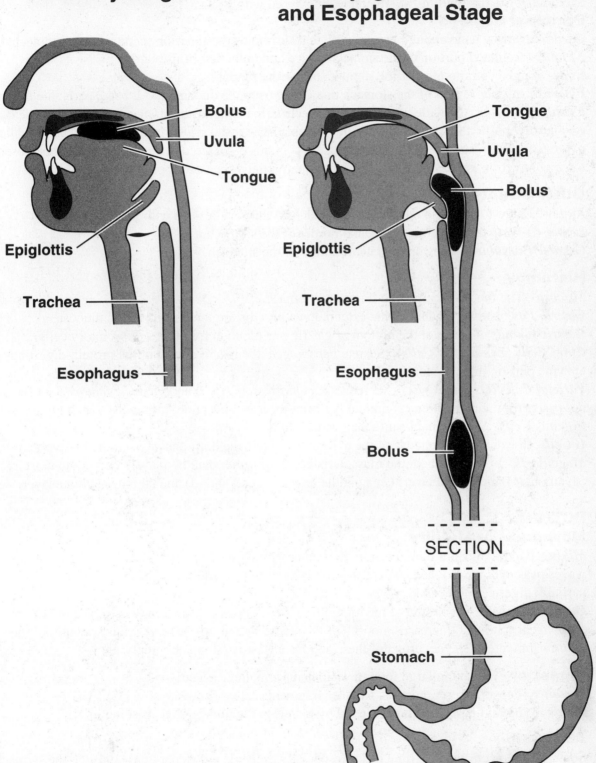

Bolus
Uvula
Tongue
Epiglottis
Trachea
Esophagus

Tongue
Uvula
Bolus
Epiglottis
Trachea
Esophagus
Bolus
SECTION
Stomach

STRUCTURES

The stomach is a J-shaped enlargement of the GI tract directly under the diaphragm. The superior portion of the stomach is a continuation of the esophagus; the inferior portion empties into the duodenum, the first part of the small intestine.

Principal Regions

Cardia   Narrow band about 2 cm in width that surrounds the superior opening of the stomach.

Fundus   Rounded portion of the stomach above and to the left of the cardia.

Body   Large central region of the stomach below the fundus.

Pylorus   Inferior region of the stomach that connects to the duodenum. It has 2 parts: the *pyloric antrum* or *antrum* (the wide portion that connects to the body of the stomach) and the *pyloric canal* (the narrow portion that leads into the duodenum). The distal region of the pyloric canal is called the *pyloric sphincter*; its wall is thicker because it contains extra circular smooth muscle; it is closed in tonic contraction, except when releasing chyme (partially digested food) into the duodenum.

Other Structures

Rugae   Large folds in the mucous lining (mucosa) that can be seen with the naked eye.

Lesser Curvature   The concave medial border of the stomach.

Greater Curvature   The convex lateral border of the stomach.

Histology

Mucosa   The mucous membrane lining the stomach.

Gastric Pits   Narrow channels that extend down into the lamina propria of the mucosa.

Gastric Glands   Glands at the bottom of gastric pits (contain four types of secretory cells) .

Chief Cells (Zymogenic Cells)   Secrete *pepsinogen*, the inactive form of the protein-digesting enzyme pepsin, into the stomach lumen.

Parietal Cells (Oxyntic Cells)   Secrete *hydrochloric acid* (HCl) and *intrinsic factor* (a substance necessary for the absorption of vitamin B_{12} from the small intestine) into the stomach lumen.

Mucous Cells   Secrete *mucus* into the stomach lumen.

G Cells   (located in the mucosa of the pylorus)   Secrete the hormone *gastrin* into the blood.

Muscularis   In the stomach, the muscularis has three rather than two layers of smooth muscle: (1) an outer longitudinal layer; (2) a middle circular layer; and (3) and an inner oblique layer.

FUNCTIONS

Movement or Motility

Mixing Waves   Several minutes after food enters the stomach, peristaltic contractions called mixing waves pass over the stomach every 20 seconds or so. This action reduces food to a solution of partially digested food called *chyme*.

Gastric Emptying   Secretion of the hormone *gastrin* causes contraction (closing) of the lower esophageal sphincter, increases motility of the stomach, and causes relaxation (opening) of the pyloric sphincter. The net effect of these actions is gastric (stomach) emptying.

Digestion   (Breaking large food molecules into smaller molecules.)

Proteins   Pepsin digests proteins, forming fragments called peptides.

Lipids   Lingual lipase, which is secreted by glands in the tongue, digests some lipids.

Absorption

Some water, electrolytes, certain drugs (especially aspirin), and alcohol pass through the stomach wall into the blood of capillaries.

STOMACH

Regions

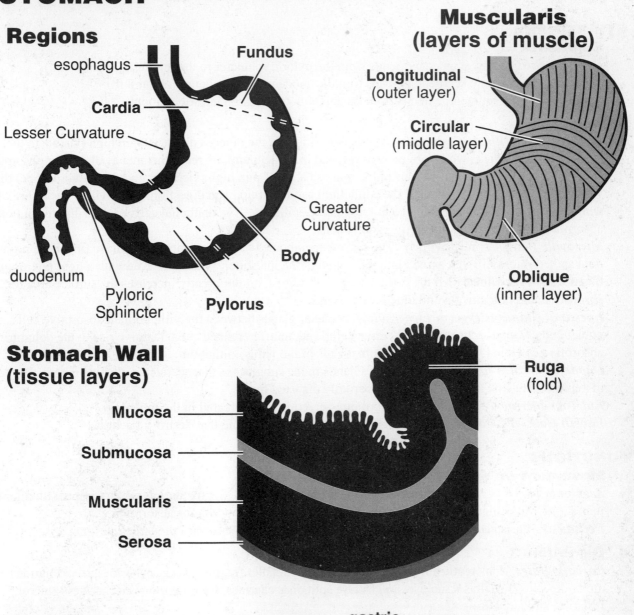

- esophagus
- **Fundus**
- **Cardia**
- Lesser Curvature
- duodenum
- Pyloric Sphincter
- **Pylorus**
- **Body**
- Greater Curvature

Muscularis (layers of muscle)

- **Longitudinal** (outer layer)
- **Circular** (middle layer)
- **Oblique** (inner layer)

Stomach Wall (tissue layers)

- **Mucosa**
- **Submucosa**
- **Muscularis**
- **Serosa**
- **Ruga** (fold)

Mucosa (stomach lining)

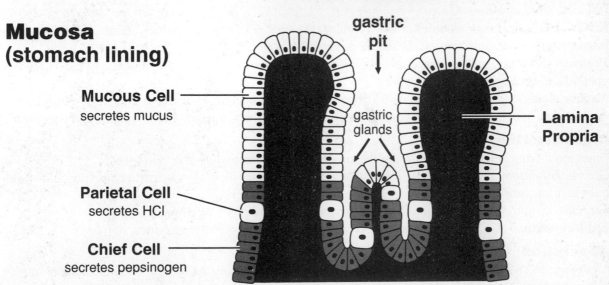

- **Mucous Cell** secretes mucus
- **Parietal Cell** secretes HCl
- **Chief Cell** secretes pepsinogen
- gastric pit
- gastric glands
- **Lamina Propria**

STRUCTURES AND FUNCTIONS / Small Intestine

STRUCTURES
Segments
Duodenum   The first segment; extends from the pyloric sphincter to the jejunum (about 10 in.).
Jejunum   The middle segment; extends fom the duodenum to the ileum (about 8 ft.).
Ileum   The final segment; extends from the jejunum to the large intestine (about 12 ft.).

Histology
Circular Folds (Plicae Circulares)   Permanent ridges in the mucosa about 10 mm high (visible to the naked eye); the plicae, which are most developed in the jejunum, increase the area available for absorption.
Villi   (singular: villus)   Projections of the mucosa about 1 mm high (visible with a light microscope) that vastly increase the surface area of the epithelium for absorption and digestion. Each villus has a core of lamina propria that contains blood and lymphatic capillaries. The epithelial cells of the villi are columnar absorptive cells and mucus-secreting goblet cells.
Microvilli   (singular: microvillus)   The plasma membranes of the absorptive cells have fingerlike projections called microvilli that extend into the lumen of the intestine. When observed with a light microscope they form a fuzzy line called the *brush border*. The brush border greatly increases the surface area for absorption and contains several digestive enzymes.
Intestinal Glands (Crypts of Lieberkühn)   Tubular glands between the villi; contain absorptive cells, goblet cells, Paneth cells, enteroendocrine cells, and undifferentiated cells that give rise to the columnar absorptive cells and mucus-secreting goblet cells of the villus epithelium.
Duodenal Glands (Brunner's Glands)   Glands in the submucosa that secrete an alkaline mucus that helps to neutralize gastric acid in the chyme (partially digested food).
Enteroendocrine Cells   Endocrine (hormone-secreting) cells located in the mucosa.
Paneth Cells   Phagocytic cells that secrete lysozyme, an enzyme that destroys bacteria.

FUNCTIONS
Movement or Motility
Segmentation   Localized contractions in areas containing food. Segmentation mixes chyme with digestive juices and brings digested food particles into contact with the mucosa for absorption.
Peristalsis   Peristalsis propels chyme through the small intestine toward the large intestine.

Digestion
Carbohydrates   Pancreatic amylase splits starch into smaller fragments (disaccharides, trisaccharides, and alpha-dextrins). Maltase, sucrase, and lactase split disaccharides, forming monosaccharides (glucose, fructose, and galactose).
Lipids   Bile salts break globules of triglycerides into smaller droplets (emulsification). Pancreatic lipase breaks triglycerides into fatty acids and monoglycerides.
Proteins   Protein-digesting enzymes (trypsin, chymotrypsin, aminopeptidase, dipeptidase, and carboxypeptidase) break proteins and peptides into amino acids.
Nucleic Acids   Ribonuclease and deoxyribonuclease digest nucleic acids (RNA and DNA), forming nucleotides. Nucleosidases and phosphatases digest nucleotides, forming bases, sugars, and phosphates.

Digestive Hormones
Enteroendocrine cells in the mucosa of the small intestine secrete digestive hormones.
Gastric Inhibitory Peptide (GIP)   Inhibits secretion of gastric juice, slows gastric emptying, and stimulates insulin secretion.
Secretin   Stimulates the release of bile and pancreatic juice rich in bicarbonate ions.
Cholecystokinin (CCK)   Stimulates the secretion of bile and pancreatic enzymes.

Absorption
Most digested food is absorbed in the small intestine. The mechanisms for absorption include : simple diffusion, facilitated diffusion, osmosis, primary active transport, and secondary active transport.

SMALL INTESTINE
Cross Section of the Small Intestine (diagramatic)

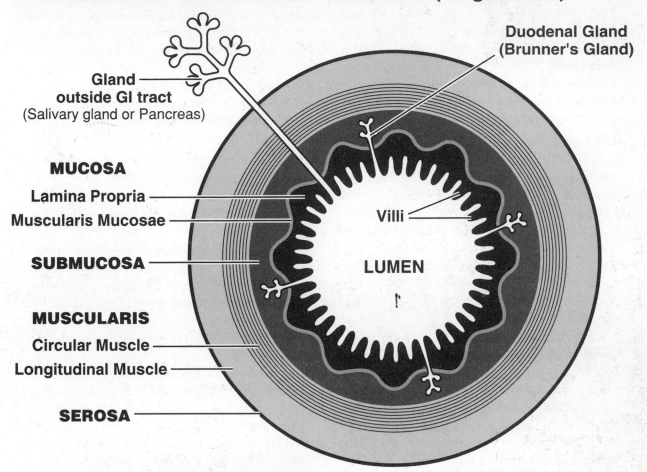

Duodenal Gland
(Brunner's Gland)

Gland
outside GI tract
(Salivary gland or Pancreas)

MUCOSA
Lamina Propria
Muscularis Mucosae

Villi

SUBMUCOSA

LUMEN

MUSCULARIS
Circular Muscle
Longitudinal Muscle

SEROSA

Lining (Mucosa and Submucosa)

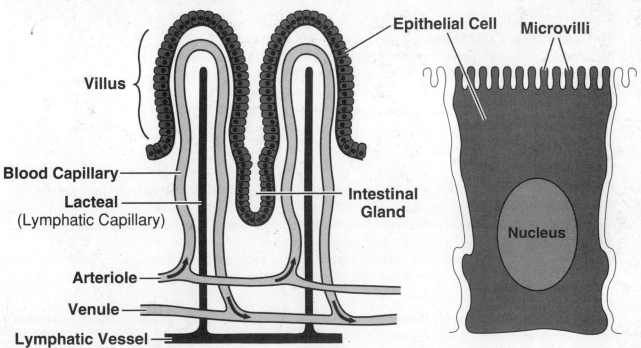

Epithelial Cell   Microvilli

Villus

Blood Capillary

Lacteal
(Lymphatic Capillary)

Intestinal
Gland

Arteriole

Venule

Lymphatic Vessel

Nucleus

STRUCTURES AND FUNCTIONS / Liver and Gallbladder

LIVER STRUCTURES

The liver is the second largest organ of the body after the skin. It is located under the diaphragm on the right side of the abdominal cavity. All nutrients absorbed by the small intestine (except complex lipids) are carried to the liver by the hepatic portal system, where they are processed for use by other organs.

Anatomy

Lobes   The liver is divided into two lobes: a large right lobe and a smaller left lobe.

Falciform Ligament   A ligament that separates the two lobes of the liver.

Round Ligament (Ligamentum Teres)   A remnant of the umbilical vein of the fetus.

Histology

Lobules   The basic structural and functional units of the liver; consist of hepatic cells (specialized epithelial cells) arranged in irregular, branching, interconnected plates around a central vein.

Hepatic Cells (Hepatocytes)   Perform a large variety of functions, including secretion of bile.

Central Vein   Vein passing through the center of a lobule; carries blood from the hepatic cells to a branch of the hepatic vein, which empties into the inferior vena cava.

Sinusoids (Sinusoidal Capillaries)   Large spaces lined with endothelium through which blood passes. They have larger diameters than other capillaries, large intercellular clefts between the endothelial cells, an incomplete basement membrane, and an irregular shape. Branches of both the hepatic portal vein (nutrient-rich deoxygenated blood) and the hepatic artery (oxygenated blood) carry blood into the liver sinusoids.

Stellate Reticuloendothelial Cells (Kupffer's Cells)   Phagocytic cells that line the sinusoids; they destroy worn-out blood cells, bacteria, and toxic substances.

Bile Ducts   *Canaliculi* (bile capillaries) are the smallest portions of the bile duct system. They are tubular spaces bordered by the plasma membranes of hepatocytes. Bile secreted by hepatocytes enters canaliculi that empty into small ducts that merge to form the right and left *hepatic ducts*; left and right hepatic ducts unite and exit the liver as the *common hepatic duct*, which joins the *cystic duct* (from the gallbladder) to form the *common bile duct*. The common bile duct and the pancreatic duct join to form a common duct called the *hepatopancreatic ampulla*, which empties into the duodenum.

LIVER FUNCTIONS

Digestive Function

Bile Production   Hepatic cells produce and secrete bile, which is used to emulsify lipids.

Nondigestive Functions

Metabolism   The liver performs many vital functions related to metabolism.

Storage   The liver stores glycogen, triglycerides, vitamins, and minerals.

Defense   Macrophages phagocytize bacteria and worn-out blood cells.

Vitamin D Activation   The skin, liver, and kidneys participate in the activation of vitamin D.

Elimination   The liver chemically alters drugs and hormones and excretes them into the bile.

GALLBLADDER

Structure

The gallbladder is a pear-shaped sac about 3 inches long. It is located in a depression (fossa) on the visceral surface of the liver.

Cystic Duct   The cystic duct carries bile from the gallbladder to the common bile duct, which is formed by the union of the cystic duct and the common hepatic duct.

Sphincter of the Hepatopancreatic Ampulla (Sphincter of Oddi)   A ring of smooth muscle that forms a valve at the hepatopancreatic ampulla (where bile enters the duodenum).

Function

The gallbladder stores and concentrates bile received from the liver. It secretes bile into the duodenum in response to stimulation by the hormone CCK (cholecystokinin).

LIVER
Hepatic Ducts

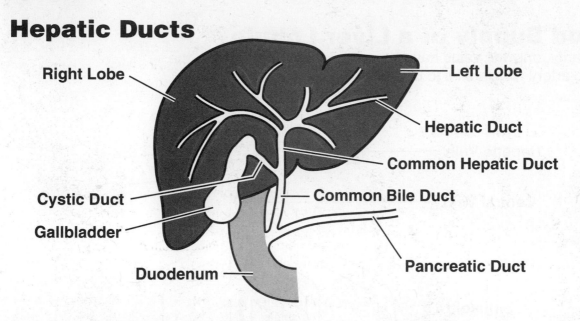

Right Lobe

Left Lobe

Hepatic Duct

Common Hepatic Duct

Cystic Duct

Common Bile Duct

Gallbladder

Pancreatic Duct

Duodenum

Liver Lobule

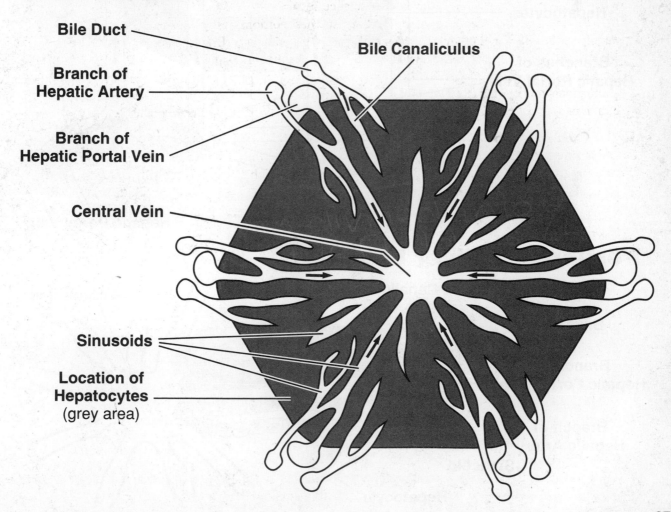

Bile Duct

Bile Canaliculus

Branch of
Hepatic Artery

Branch of
Hepatic Portal Vein

Central Vein

Sinusoids

Location of
Hepatocytes
(grey area)

LIVER LOBULE

Blood Supply of a Liver Lobule

(For clarity, only the veins have been illustrated. Branches of the hepatic artery run parallel to branches of the hepatic portal vein.)

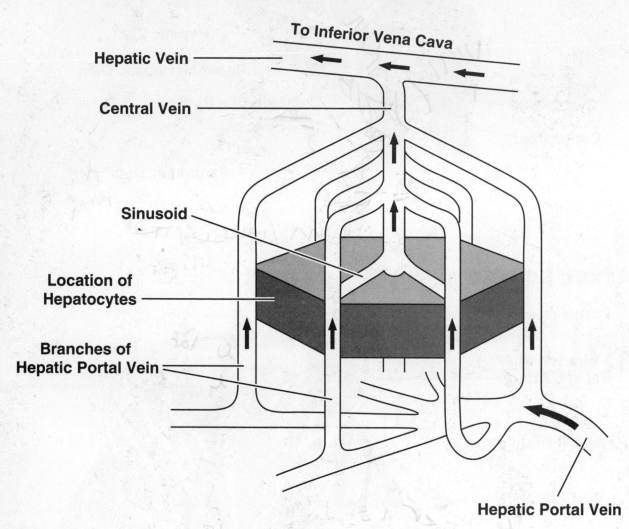

To Inferior Vena Cava

Hepatic Vein

Central Vein

Sinusoid

Location of Hepatocytes

Branches of Hepatic Portal Vein

Hepatic Portal Vein

Portion of a Liver Lobule

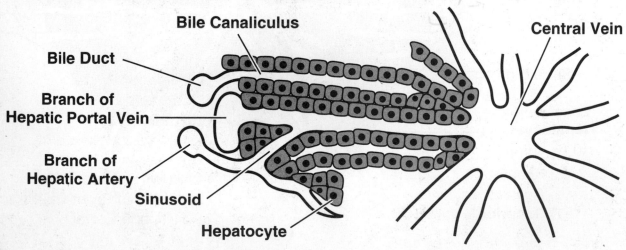

Bile Canaliculus

Central Vein

Bile Duct

Branch of Hepatic Portal Vein

Branch of Hepatic Artery

Sinusoid

Hepatocyte

HEPATIC PORTAL SYSTEM

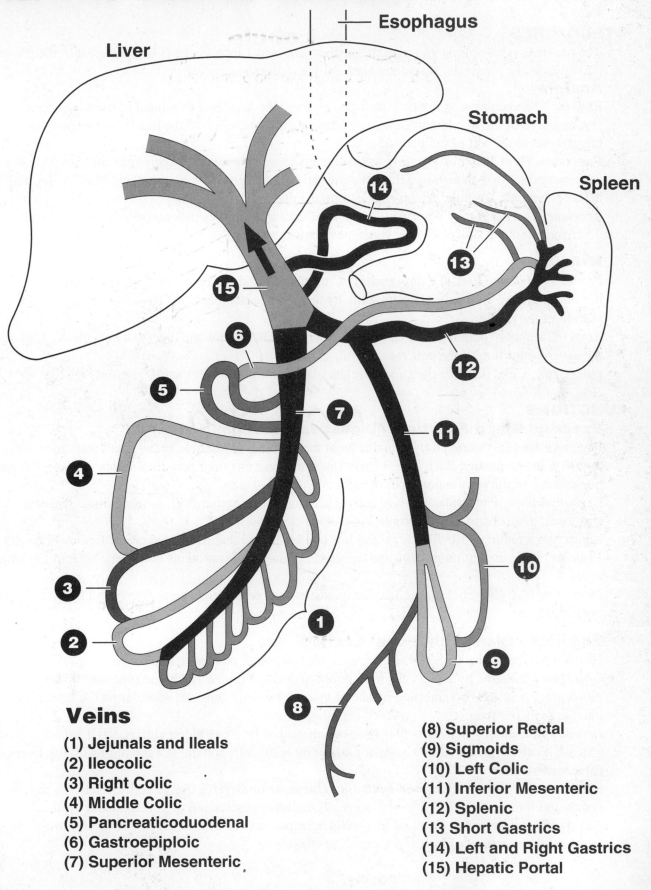

Esophagus

Liver

Stomach

Spleen

Veins

(1) Jejunals and Ileals
(2) Ileocolic
(3) Right Colic
(4) Middle Colic
(5) Pancreaticoduodenal
(6) Gastroepiploic
(7) Superior Mesenteric

(8) Superior Rectal
(9) Sigmoids
(10) Left Colic
(11) Inferior Mesenteric
(12) Splenic
(13 Short Gastrics
(14) Left and Right Gastrics
(15) Hepatic Portal

STRUCTURES AND FUNCTIONS / Pancreas

STRUCTURES

The pancreas is an oblong gland about 5 inches long and 1 inch thick. It lies posterior to the greater curvature of the stomach. It secretes hormones and digestive juices.

Anatomy

Regions   The pancreas is divided into three regions: the *head* (the expanded portion near the C-shaped curve of the duodenum); the *body* (superior and to the left of the head); and the *tail* (the tapered end to the left of the body).

Pancreatic Duct (Duct of Wirsung)   Small ducts carrying pancreatic secretions unite to form the pancreatic duct, which usually joins the common bile duct from the liver and gallbladder, forming the hepatopancreatic ampulla, which empties into the duodenum.

Accessory Duct (Duct of Santorini)   A smaller duct that empties pancreatic secretions into the duodenum about 1 inch above the hepatopancreatic ampulla.

Histology

Pancreatic Islets (Islets of Langerhans)   Clusters of glandular epithelial cells that secrete the hormones glucagon, insulin, somatostatin, and pancreatic polypeptide. They form the *endocrine* portion of the pancreas.

Acini   Clusters of acinar cells that secrete pancreatic juice (fluid and digestive enzymes). They form the *exocrine* portion of the pancreas.

Duct Cells   Cells that line the pancreatic ducts secrete fluid rich in bicarbonate ions (HCO_3^-).

FUNCTIONS

Exocrine Gland Functions (Digestive Functions)   acini

Digestive enzymes secreted by acinar cells of the pancreas are carried to the small intestine (duodenum) via the pancreatic duct. In the small intestine these enzymes play important roles in the digestion of each of the four major types of food.

Carbohydrates   Pancreatic amylase digests starches (polysaccharides), forming disaccharides (maltose), trisaccharides (maltotriose), and alpha-dextrins.

Lipids   Pancreatic lipase digests neutral fats (triglycerides) that have been emulsified by bile salts.

Proteins   Trypsin, chymotrypsin, and carboxypeptidase digest proteins and peptides, forming amino acids.

Nucleic Acids   Ribonuclease and deoxyribonuclease digest nucleic acids (RNA and DNA), forming nucleotides.

Endocrine Gland Functions   pancreatic islets

Hormones secreted by cells of the pancreatic islets regulate metabolic functions.

Glucagon   Secreted by alpha cells in response to decreased levels of blood glucose. It increases blood glucose levels by inhibiting uptake of glucose by body cells and stimulating the release of glucose from the liver.

Insulin   Secreted by beta cells in response to increased levels of blood glucose. It decreases blood glucose levels by stimulating uptake of glucose by body cells and the storage of glucose by liver cells (glycogen synthesis).

Somatostatin (Growth Hormone-Inhibiting Hormone or GHIH)   Secreted by delta cells in response to increased levels of blood glucose. It inhibits the secretion of glucagon and insulin.

Pancreatic Polypeptide   Secreted by F-cells in response to decreased levels of blood glucose. It regulates the secretion of pancreatic digestive enzymes.

PANCREAS

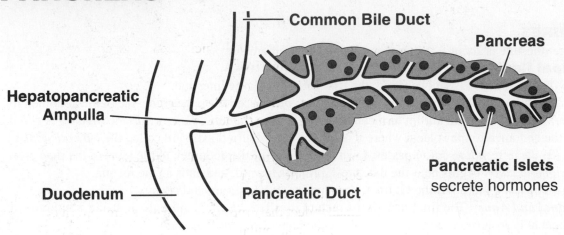

Common Bile Duct

Pancreas

Hepatopancreatic Ampulla

Pancreatic Islets
secrete hormones

Duodenum

Pancreatic Duct

Exocrine Portion

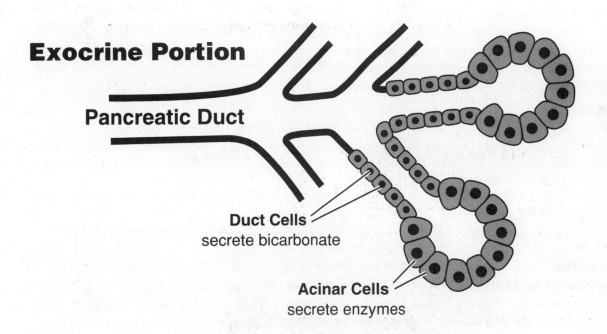

Pancreatic Duct

Duct Cells
secrete bicarbonate

Acinar Cells
secrete enzymes

Endocrine Portion

Pancreatic Islet

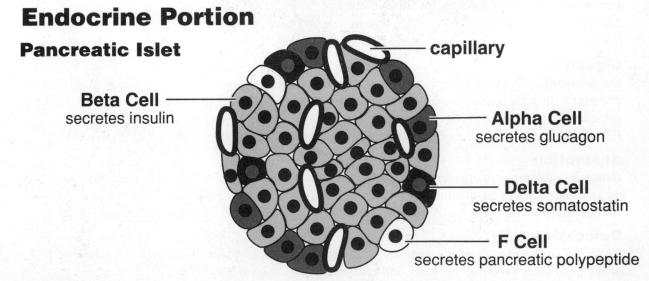

capillary

Beta Cell
secretes insulin

Alpha Cell
secretes glucagon

Delta Cell
secretes somatostatin

F Cell
secretes pancreatic polypeptide

STRUCTURES AND FUNCTIONS / Large Intestine

STRUCTURES
The large intestine is about 5 feet long and 2.5 inches in diameter.

Principal Regions
Cecum A pouch about 2.5 inches long; the first part of the large intestine.

Colon The colon has 4 segments: (1) the *ascending colon* ascends on the right side of the abdomen to the liver, where it makes an abrupt turn called the *right colic (hepatic) flexure*; (2) the *transverse colon* crosses the abdomen to the spleen, where it makes an abrupt turn downward called the *left colic (splenic) flexure*; (3) the *descending colon* passes down the left side of the abdomen to the level of the iliac crest; (4) the *sigmoid colon* begins near the iliac crest, projects inward, and ends as the rectum.

Rectum The last 8 inches of the GI tract; lies anterior to the sacrum and coccyx.

Anal Canal and Anus The final inch of the rectum is called the *anal canal*; the mucous membrane of the anal canal is arranged in longitudinal folds called *anal columns* that contain a network of arteries and veins. The opening of the anal canal to the exterior is called the *anus*. An *internal sphincter* (involuntary smooth muscle) and an *external sphincter* (voluntary skeletal muscle) control defecation.

Other Structures
Mesocolon Peritoneum that attaches the large intestine to the posterior abdominal wall.

Ileocecal Sphincter (Valve) A fold of mucous membrane that acts as a valve, regulating the passage of materials from the ileum into the large intestine (at the cecum).

Vermiform Appendix A twisted, closed-ended tube about 3 inches long attached to the cecum.

Histology
Mucosa No permanent folds (rugae) or villi are found in the lining (mucosa) of the large intestine.

Epithelium Consists of columnar epithelium (for water absorption) and mucus-secreting goblet cells.

Taeniae Coli (singular: taenia coli) Thickened portions of longitudinal muscles, forming three bands that run the length of most of the large intestine.

Haustra (singular: haustrum) Pouches which give the colon a puckered appearance; formed by tonic contractions of the taeniae coli.

Epiploic Appendages Small fat-filled pouches of visceral peritoneum attached to the taeniae coli.

FUNCTIONS

Movement or Motility
Haustral Churning Haustra remain relaxed and distended while they fill up; when stretched to a certain degree they contract and squeeze their contents into the next haustra.

Peristalsis Occurs at a slower rate than in other portions of the GI tract.

Mass Peristalsis A strong peristaltic wave initiated by food in the stomach; it begins at the middle of the transverse colon and drives the contents of the colon into the rectum.

Digestion
Carbohydrates Bacteria ferment carbohydrates and release hydrogen, carbon dioxide, and methane gas.

Proteins Bacteria convert remaining proteins to amino acids and break down amino acids into simpler substances: indole, skatole, hydrogen sulfide, and fatty acids.

Bilirubin Bacteria decompose bilirubin to simpler pigments, which give feces their brown color.

Absorption
Osmosis Although most water is absorbed in the small intestine, the large intestine absorbs enough to make it important in maintaining the body's water balance. The absorption of water occurs passively following the active transport of sodium. Electrolytes and water-soluble vitamins are also absorbed.

Defecation
Feces Feces consist of water, inorganic salts, sloughed off epithelial cells from the GI tract, bacteria, products of bacterial decomposition, and undigested parts of food.

LARGE INTESTINE
Anatomy

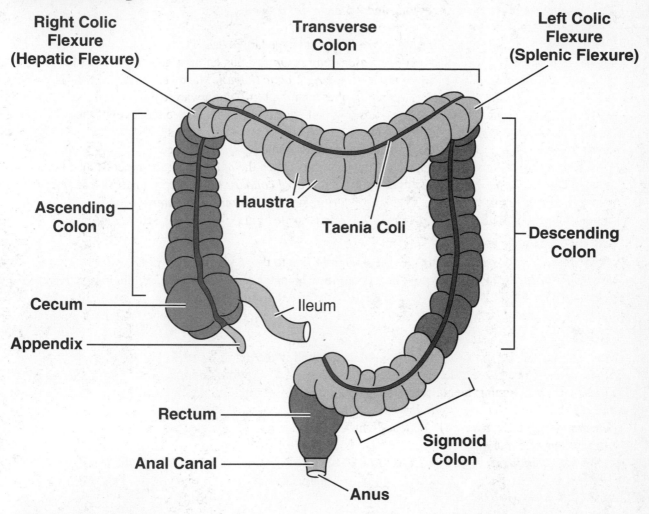

Right Colic Flexure (Hepatic Flexure)

Transverse Colon

Left Colic Flexure (Splenic Flexure)

Ascending Colon

Haustra

Taenia Coli

Descending Colon

Cecum

Ileum

Appendix

Rectum

Sigmoid Colon

Anal Canal

Anus

Histology

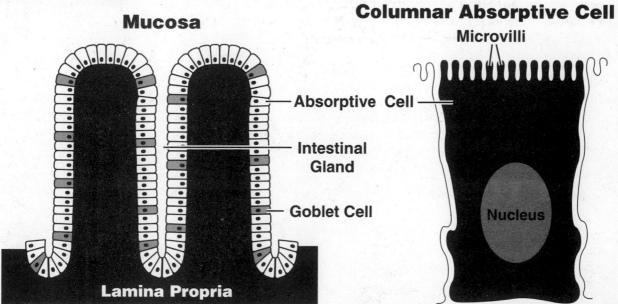

Mucosa

Columnar Absorptive Cell

Microvilli

Absorptive Cell

Intestinal Gland

Goblet Cell

Nucleus

Lamina Propria

3 Digestion and Absorption

DIGESTION AND ABSORPTION / Overview

DIGESTION

Digestion is the breakdown of food by both mechanical and chemical processes.

Mechanical Digestion

Mechanical digestion refers to various movements of the GI tract.

Mouth   In the mouth food is reduced to a soft, flexible mass called a *bolus* through the grinding action of the teeth (chewing or *mastication*) and the mixing of the food with saliva by the tongue.

Esophagus   The bolus is pushed through the esophagus to the stomach by waves of involuntary contractions called *peristalsis*.

Stomach   In the stomach, rippling, peristaltic movements (called *mixing waves*) mix the food with gastric juice and reduce it to a thin liquid called *chyme*. Each mixing wave forces a small amount of the stomach contents into the duodenum through the pyloric sphincter.

Small Intestine   In the small intestine, localized contractions in areas containing food are called *segmentation*. Segmentation mixes chyme with digestive juices and brings the digested particles of food into contact with the absorptive epithelial cells lining the intestine. *Peristalsis* propels the chyme onward through the intestinal tract.

Large Intestine   In the large intestine, there are three types of mechanical digestion: (1) *haustral churning* moves the contents of the large intestine from haustrum (pouch) to haustrum; (2) *mass peristalsis* is a strong peristaltic wave that drives the contents of the colon into the rectum; (3) *peristalsis* also occurs.

Chemical Digestion

Chemical digestion refers to the breakdown of large molecules of food into smaller molecules that can be absorbed and used by body cells. These products of digestion are small enough to pass through the epithelial cells lining the GI tract and enter the blood (or lymph).

Hydrolysis (*hydro* = water; *lysis* = to split)   Hydrolysis means to split apart by using water. Large molecules such as carbohydrates, lipids, and proteins are broken down into smaller molecules by *hydrolytic reactions*. These reactions are catalyzed by digestive enzymes. To function, an enzyme must bind with its *substrate* (a substance in an enzyme-mediated reaction), forming an enzyme-substrate complex. Enzymes are generally named by adding the suffix *-ase* to the name of the substrate (the enzyme for the substrate sucr*ose* is sucr*ase*). After binding with its enzyme, a substrate breaks apart into the products of the reaction. A single enzyme can be used repeatedly, converting about 100,000 substrate molecules to products in one second.

Carbohydrates   The final products of carbohydrate digestion are monosaccharides.

Lipids   The final products of lipid digestion are monoglycerides and fatty acids.

Proteins   The final products of protein digestion are amino acids.

ABSORPTION

Absorption is the passage of digested food from the GI tract into the blood (or lymph). About 90% of all absorption of nutrients takes place in the small intestine; the other 10% occurs in the stomach and large intestine. The microvilli of the epithelial cells lining the small intestine and large intestine greatly increase the total surface area available for absorption.

Monosaccharides   Move by secondary active transport or facilitated diffusion into epithelial cells.

Monoglycerides and Fatty Acids   Move by simple diffusion into epithelial cells.

Amino Acids   Move by secondary active transport into epithelial cells.

Vitamins, Minerals, and Water   Most vitamins are absorbed by simple diffusion. Minerals are absorbed by several mechanisms (simple diffusion, active transport, and secondary active transport). All water absorption in the GI tract occurs by osmosis.

ENZYME ACTION

Example : hydrolysis of sucrose (cane sugar).
The action of sucrase (enzyme) on sucrose (substrate).

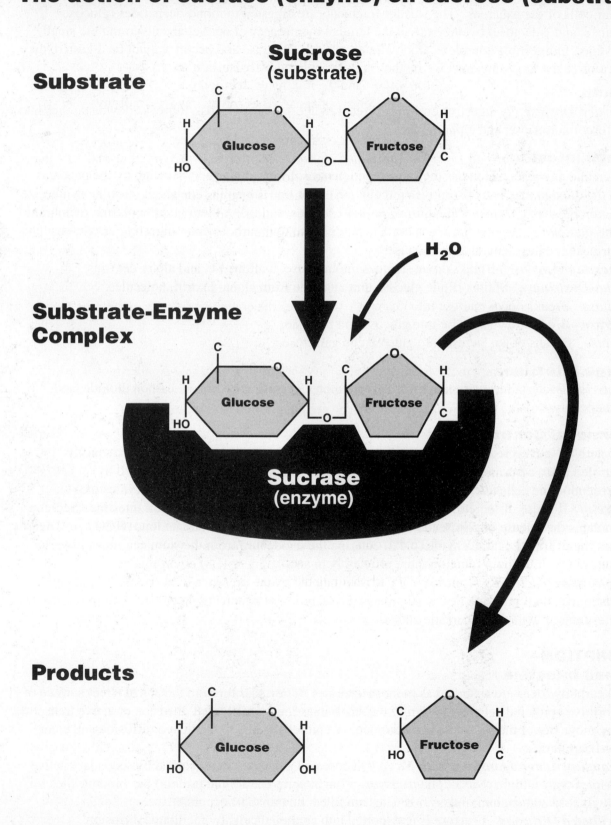

Substrate

Sucrose
(substrate)

Substrate-Enzyme Complex

H_2O

Sucrase
(enzyme)

Products

DIGESTION

Starches are split into smaller fragments by amylases that are secreted by the salivary glands and the acinar cells of the pancreas. The smaller fragments are digested to monosaccharides (glucose, fructose, and galactose) by enzymes in the luminal membranes of epithelial cells lining the small intestine. Indigestible complex carbohydrates, such as cellulose and pectin, cannot be digested; they move on to the large intestine, where they are metabolized by bacteria.

Mouth

Salivary Amylase   A digestive enzyme secreted by the salivary glands. It breaks down starches into maltose, maltotriose, and alpha-dextrins.

Small Intestine

Pancreatic amylase, which has the same functions as salivary amylase, is secreted by the pancreas into the duodenum. The absorptive epithelial cells that line the villi of the small intestine synthesize several digestive enzymes, called *brush border enzymes*, and insert them into the plasma membranes of the microvilli. Among the brush border enzymes are four carbohydrate-digesting enzymes: alpha-dextrinase, maltase, sucrase, and lactase.

Pancreatic Amylase   Breaks down starches into maltose, maltotriose, and alpha-dextrins.

Alpha-Dextrinase   Splits off one glucose unit at a time from alpha-dextrin molecules.

Maltase   Breaks down maltose into glucose.

Sucrase   Breaks down sucrose into glucose and fructose.

Lactase   Breaks down lactose into glucose and galactose.

Large Intestine

Bacteria   Bacteria ferment undigested carbohydrates, releasing hydrogen, carbon dioxide, and methane gas.

Hormonal Control

The activity and secretion of the pancreatic enzymes that break down carbohydrates is partly controled by two hormones, secretin and cholecystokinin (CCK), which are secreted by enteroendocrine cells that are scattered among the epithelial cells lining the small intestine.

Secretin   Release of secretion is stimulated by the presence of acid in the small intestine. Secretin stimulates the release of pancreatic juice and bile that are rich in bicarbonate ions (HCO_3^-). This raises the pH of the acidic chyme (pH 2) entering the duodenum from the stomach; digestive enzymes in the duodenum function more efficiently in an alkaline (pH 8) environment.

Cholecystokinin (CCK)   Release of CCK is stimulated by the presence of amino acids and fatty acids in the small intestine. CCK stimulates the secretion of pancreatic juice rich in digestive enzymes, one of which is pancreatic amylase.

ABSORPTION

Small Intestine

All carbohydrates are absorbed as monosaccharides. They pass through the apical (free) surface of absorptive epithelial cells in the villi of the small intestine by facilitated diffusion or active transport. They move out of the sides and basal surfaces of epithelial cells by facilitated diffusion and enter blood capillaries.

Secondary Active Transport (with Na⁺)   Glucose and galactose are absorbed by secondary active transport with sodium (Na^+). The *transporter* (an integral membrane protein) has binding sites for both glucose and sodium; unless both sites are filled, no transport occurs.

Facilitated Diffusion   Fructose is transported into epithelial cells by facilitated diffusion.

CARBOHYDRATES

Digestion
Final step in carbohydrate digestion : disaccharides are split, forming monosaccharides.

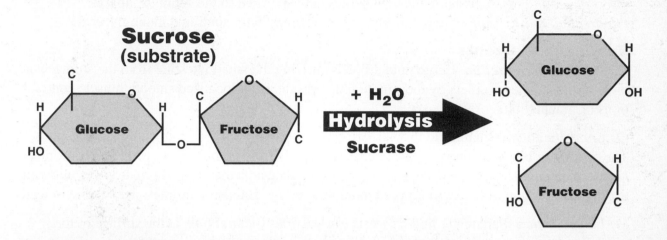

Absorption
Monosaccharides move by diffusion or active transport from the lumen of the small intestine into epithelial cells and then into the blood.

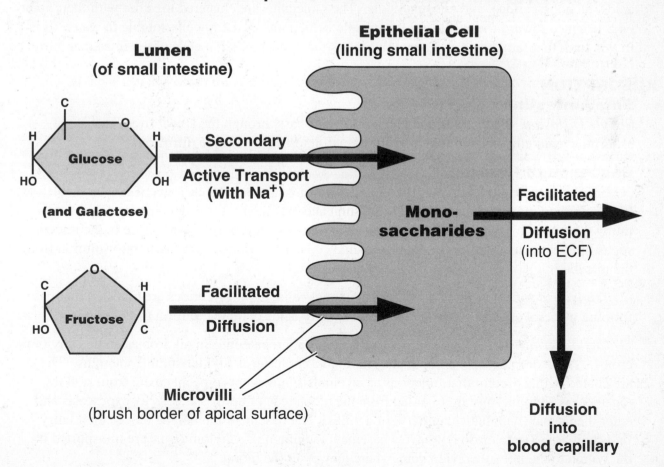

DIGESTION
Small Intestine

Bile   Bile is produced by the liver and stored in the gallbladder. When stimulated by the hormone *cholecystokinin (CCK)*, the gallbladder contracts, ejecting bile into the duodenum. Bile salts emulsify lipids, increasing the total surface area exposed to the action of lipase.

Pancreatic Lipase   Lipase digests triglycerides, forming fatty acids and monoglycerides.

Sequence of Events

(1) Mechanical Agitation   Large lipid droplets entering the small intestine from the stomach are broken into smaller droplets by the agitation that results from localized smooth muscle contractions (segmentation).

(2) Emulsification   Small lipid droplets are coated with bile salts, forming emulsion droplets. The coating of bile salts prevents the emulsion droplets from combining to form larger droplets. Many small emulsion droplets have a larger total surface area than large lipid droplets, allowing the digestive action of lipase to proceed more efficiently. Emulsion droplets are soluble in water.

(3) Lipase Action   Pancreatic lipase digests triglycerides (neutral fats) at the surface of the emulsion droplets, forming fatty acids and monoglycerides, which are insoluble in water.

(4) Micelles   The fatty acids and monoglycerides combine with bile salts, forming very small, water-soluble aggregates called micelles. Micelles are similar in structure to emulsion droplets, but much smaller (2 – 10 nm in diameter). They consist of bile salts (20 – 50 molecules), fatty acids, monoglycerides, and phospholipids. The molecules are clustered together with their polar ends oriented toward the surface of the micelle, which makes the micelle soluble in water. It is in this form that fatty acids and monoglycerides reach the epithelial cells of the intestinal villi.

ABSORPTION
Small Intestine

Simple Diffusion   Fatty acids and monoglycerides pass through the apical (free) surface of absorptive epithelial cells in the villi of the small intestine by simple diffusion.

Sequence of Events

(1) Equilibrium and Diffusion   Although lipids are very insoluble in water, a few molecules of fatty acids and monoglycerides exist in solution and are free to diffuse into the epithelial cells lining the GI tract. The lipid molecules in solution are in equilibrium with those in the micelle; so, as lipid molecules diffuse into the epithelial cells, a few more are released into solution from the micelle.

(2) Triglycerides Resynthesized   Within epithelial cells, monoglycerides are digested by lipase, forming glycerol and fatty acids. Glycerol and fatty acids combine to form triglycerides.

(3) Chylomicrons   The resynthesized triglycerides aggregate into small droplets called chylomicrons. These droplets also contain phospholipids, cholesterol, and fat-soluble vitamins. They are coated with a protein that functions as an emulsifying agent (prevents them from sticking together). Chylomicrons are released from the blood side of epithelial cells by exocytosis and enter the lacteal (lymphatic capillary) of a villus. The basement membrane blocks their entry into capillaries (lacteals do not have a basement membrane). Chylomicrons are transported in the lymph to the left subclavian vein, where they enter the blood.

LIPIDS

Digestion

Large fat droplets must be broken down into smaller droplets and coated with bile salts (emulsification) before they can be efficiently digested by the enzyme lipase. Lipase splits two fatty acids from a triglyceride, forming 1 monoglyceride and 2 fatty acids.

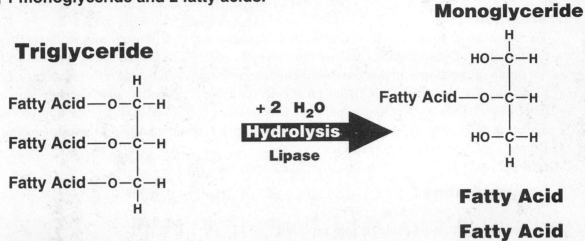

Triglyceride

Fatty Acid—O—C—H

Fatty Acid—O—C—H

Fatty Acid—O—C—H

+ 2 H₂O
Hydrolysis
Lipase

Monoglyceride

HO—C—H

Fatty Acid—O—C—H

HO—C—H

Fatty Acid

Fatty Acid

Absorption

Fatty acids and monoglycerides combine with bile salts, forming tiny water-soluble spheres called micelles. As they are released from the micelles, fatty acids and monoglycerides move by simple diffusion from the lumen of the small intesine into epithelial cells.

Inside the epithelial cells triglycerides are resynthesized, forming spherical masses called chylomicrons, which leave the epithelial cells by exocytosis and enter lymphatic capillaries (lacteals).

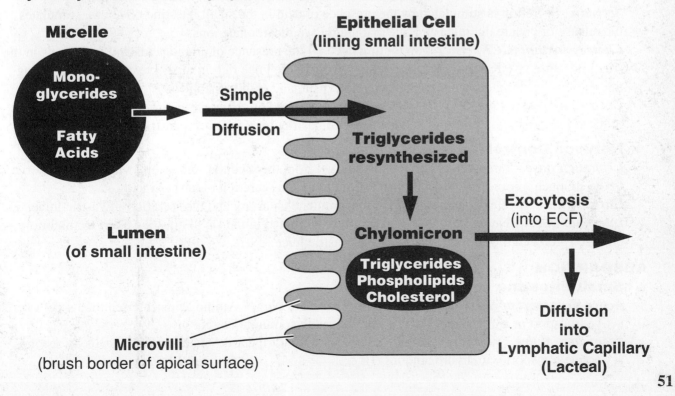

Micelle

Epithelial Cell
(lining small intestine)

Mono-glycerides

Fatty Acids

Simple

Diffusion

Triglycerides resynthesized

Exocytosis
(into ECF)

Lumen
(of small intestine)

Chylomicron

Triglycerides
Phospholipids
Cholesterol

**Diffusion into Lymphatic Capillary
(Lacteal)**

Microvilli
(brush border of apical surface)

51

DIGESTION

Proteins are broken down into small peptides by the enzyme *pepsin* in the stomach and by the pancreatic enzymes *trypsin* and *chymotrypsin* in the small intestine. The peptides are broken down into amino acids by the pancreatic enzyme *carboxypeptidase* and the brush border enzymes *aminopeptidase* and *dipeptidase*.

Stomach

The hormone *gastrin*, which is secreted by cells in the stomach in response to stretching, promotes the secretion of gastric juice. *Gastric juice* consists of the secretions of three types of cells found in the lining of the stomach: *mucous cells*, which secrete mucus; *parietal cells* (or oxyntic cells), which secrete hydrochloric acid (HCl); and *chief cells* (or zymogenic cells), which secrete pepsinogen (the inactive precursor of the enzyme pepsin). Mucus coats the lining of the stomach, protecting the epithelial cells from the damaging effects of HCl (pH 2) and the protein-digesting enzyme pepsin.

Pepsin   Splits proteins into peptides.

Hydrochloric Acid   Converts inactive pepsinogen into pepsin.

Small Intestine

Trypsin   Splits proteins into peptides.

Chymotrypsin   Splits proteins into peptides.

Carboxypeptidase   Splits off the terminal amino acid from the carboxyl end of a peptide.

Aminopeptidase   Splits off the terminal amino acid from the amino end of a peptide.

Dipeptidase   Splits dipeptides, forming amino acids.

Large Intestine

Bacteria   Bacteria convert undigested proteins into amino acids and break down amino acids into simpler substances (indole, skatole, hydrogen sulfide, and fatty acids).

Hormonal Control

Gastrin   Secretion is stimulated by distension of the stomach and the presence of peptides. Gastrin stimulates the secretion of gastric juice and increases gastric motility (mixing waves).

Secretin   Secretion is stimulated by the presence of acid in the small intestine. Secretin stimulates the release of pancreatic juice and bile that are rich in bicarbonate ions.

Cholecystokinin (CCK)   Secretion is stimulated by the presence of amino acids and fatty acids in the small intestine. CCK stimulates the release of pancreatic juice rich in digestive enzymes, including the protein-digesting enzymes trypsin, chymotrypsin, and carboxypeptidase.

Gastric Inhibitory Peptide (GIP)   Secretion is stimulated by the presence of fatty acids and glucose in the small intestine. GIP inhibits the release of gastric juice and slows gastric emptying.

Nervous Control

Cephalic Phase   Parasympathetic impulses travel from the medulla (brainstem) via the vagus nerve to the stomach, promoting peristalsis and stimulating the secretion of gastric juice.

Gastric Phase   Reflexes within the stomach stimulate churning and the secretion of gastric juice.

Intestinal Phase   Neural (and hormonal) reflexes initiated in the small intestine exert an inhibitory effect on stomach motility and the release of gastric juice.

ABSORPTION

Small Intestine

Active Transport or Secondary Active Transport (with Na$^+$)   Amino acids enter epithelial cells by active transport or by secondary active transport with sodium ions (Na$^+$).

Secondary Active Transport (with H$^+$)   Dipeptides and tripeptides enter epithelial cells by secondary active transport with hydrogen ions (H$^+$).

PROTEINS

Digestion

Final step in protein digestion : dipeptides are split, forming amino acids; or amino acids are cleaved from the carboxyl or amino end of a peptide.

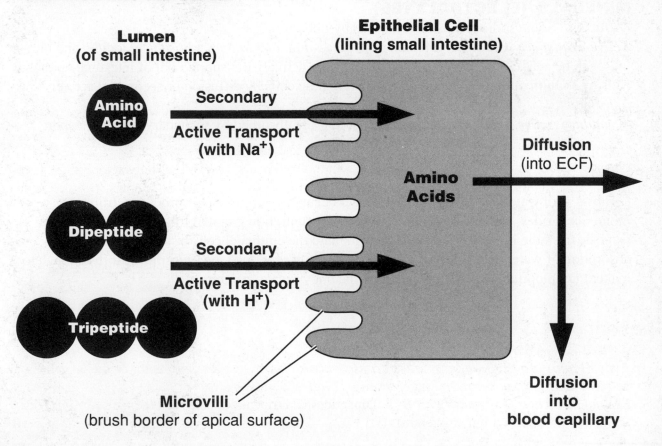

Dipeptide (substrate)

+ H_2O

Hydrolysis

Dipeptidase

Glycine

Alanine

Absorption

Amino acids, dipeptides, and tripeptides move by secondary active transport from the lumen of the small intestine into epithelial cells and then into the blood.

Lumen (of small intestine)

Epithelial Cell (lining small intestine)

Amino Acid

Secondary Active Transport (with Na^+)

Amino Acids

Diffusion (into ECF)

Dipeptide

Secondary Active Transport (with H^+)

Tripeptide

Microvilli (brush border of apical surface)

Diffusion into blood capillary

DIGESTION AND ABSORPTION / Vitamins, Minerals, and Water

DIGESTION

Vitamins, minerals, and water do not have to be digested (broken down into smaller molecules). Vitamins are small organic molecules — small enough to pass through the membranes of epithelial cells that line the GI tract. Minerals are inorganic substances that break apart into ions (charged particles) when dissolved in water. They are called electrolytes because they are able to conduct electricity. Most ions are charged atoms, the smallest nutrients that are absorbed by the GI tract. Water, of course, is a small inorganic molecule that passes easily through the membranes of epithelial cells.

ABSORPTION
VITAMINS
Water-Soluble Vitamins (B Vitamins and Vitamin C)

Diffusion   Most water-soluble vitamins are absorbed by diffusion.
Active Transport   Vitamin B_{12} combines with a substance called intrinsic factor, which is produced by the stomach. The combination is absorbed by active transport.

Fat-Soluble Vitamins (Vitamins A, D, E, and K)

Lipid Absorption   Fat-soluble vitamins are absorbed by the same mechanism that is used for lipids. They are included along with triglycerides in micelles.

MINERALS (ELECTROLYTES)

Diffusion   Some sodium ions are absorbed by diffusion.
Active Transport   Chloride, calcium, iron, potassium, magnesium, phosphate, iodide, and nitrate ions, can be actively transported. Sodium ions are actively transported out of intestinal epithelial cells by sodium pumps after they have entered these cells by diffusion or secondary active transport.
Secondary Active Transport   Some sodium ions are absorbed by secondary active transport.

WATER

Osmosis   All water absorption in the GI tract occurs by osmosis. Absorbed electrolytes, monosaccharides, and amino acids establish a concentration gradient for water that promotes osmosis from the lumen of the small intestine into the epithelial cells of the GI tract. Most absorption of water occurs in the small intestine; however, the large intestine absorbs enough to make it an important organ in maintaining the body's water balance.

VITAMINS, MINERALS, and WATER

Digestion
Vitamins are small organic molecules; minerals are inorganic molecules that break apart into ions when dissolved in water. Vitamins and minerals do not have to be digested, since they are already small enough to pass through the epithelial cells that line the small intestine.

Absorption
Water-soluble vitamins are absorbed by simple diffusion.
Fat-soluble vitamins are absorbed by the same mechanism as lipids.
Minerals are absorbed by diffusion or active transport.
Most water absorption occurs in the small intestine by osmosis.

Fluid Balance in the Gastrointestinal Tract

Fluids Ingested and Secreted   Fluids Absorbed

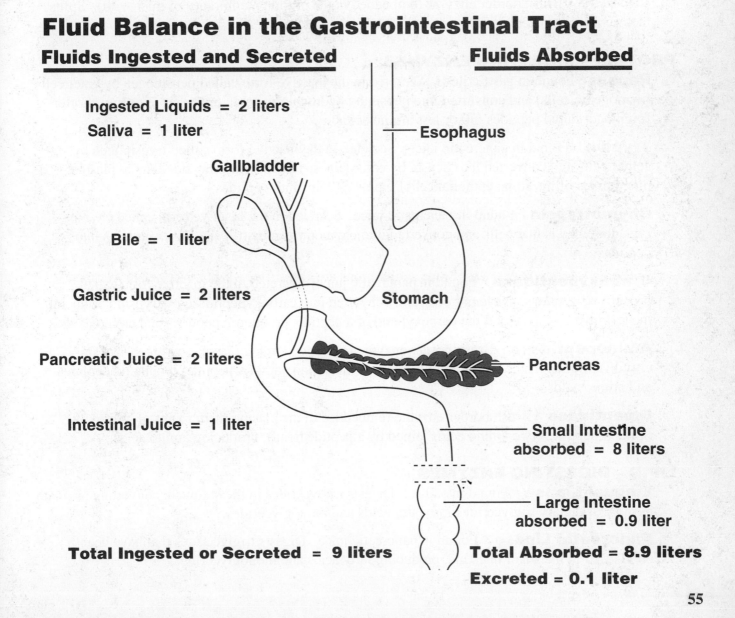

Ingested Liquids = 2 liters
Saliva = 1 liter

Esophagus

Gallbladder

Bile = 1 liter

Gastric Juice = 2 liters

Stomach

Pancreatic Juice = 2 liters

Pancreas

Intestinal Juice = 1 liter

Small Intestine absorbed = 8 liters

Large Intestine absorbed = 0.9 liter

Total Ingested or Secreted = 9 liters   **Total Absorbed = 8.9 liters**

Excreted = 0.1 liter

CARBOHYDRATE – DIGESTING ENZYMES

Salivary Amylase   Found in saliva. Splits starches (long-chain polysaccharides) into smaller fragments: the disaccharide maltose; the trisaccharide maltotriose; and short-chain glucose polymers called alpha-dextrins.

Pancreatic Amylase   Found in pancreatic juice. It has the same function as salivary amylase.

Alpha – Dextrinase   Brush border enzyme (embedded in the plasma membranes of microvilli). Acts on alpha-dextrins, clipping off one glucose unit at a time.

Maltase   Brush border enzyme (embedded in the plasma membranes of microvilli). Splits maltose or maltotriose into two or three molecules of glucose, respectively.

Sucrase   Brush border enzyme (embedded in the plasma membranes of microvilli). Splits sucrose into a molecule of glucose and a molecule of fructose.

Lactase   Brush border enzyme (embedded in the plasma membranes of microvilli). Splits lactose into a molecule of glucose and a molecule of galactose.

PROTEIN – DIGESTING ENZYMES

Pepsin   Found in gastric juice. Secreted in the inactive form called pepsinogen by chief cells (zymogenic cells) and converted into pepsin by hydrochloric acid. Splits proteins into shorter fragments called peptides (50 or less amino acids).

Trypsin   Found in pancreatic juice. Secreted in the inactive form called trypsinogen by acinar cells and converted into pepsin by enterokinase (an enzyme embedded in the plasma membranes of intestinal epthelial cells). Splits proteins into peptides.

Chymotrypsin   Found in pancreatic juice. Secreted in the inactive form called chymotrypsinogen by acinar cells and converted into chymotrypsin by trypsin. Splits proteins into peptides.

Carboxypeptidase   Found in pancreatic juice. Secreted in the inactive form called procarboxypeptidase by acinar cells and converted into carboxypeptidase by trypsin. Splits off the terminal amino acid at the carboxyl end of a peptide, producing peptides and amino acids.

Aminopeptidase   Brush border enzyme (embedded in the plasma membranes of microvilli). Splits off the terminal amino acid at the amino end of a peptide, producing peptides and amino acids.

Dipeptidase   Brush border enzyme (embedded in the plasma membranes of microvilli). Splits dipeptides (two amino acids joined by a peptide bond), producing amino acids.

LIPID – DIGESTING ENZYMES

Lingual Lipase   Found in saliva. Digests triglycerides in the stomach, converting as much as 30% of dietary triglycerides into fatty acids and monoglycerides.

Pancreatic Lipase   Found in pancreatic juice. Digests triglycerides that have been emulsified in the small intestine, producing fatty acids and monoglycerides.

DIGESTIVE ENZYMES

Enzyme	Source	Substrate	Products
CARBOHYDRATE – DIGESTING ENZYMES			
Salivary Amylase	Salivary Glands	Starches	Alpha-Dextrins Maltose Maltotriose
Pancreatic Amylase	Acinar Cells (Pancreas)	Starches	Alpha-Dextrins Maltose Maltotriose
Alpha-Dextrinase	Brush Border (Small Intestine)	Alpha-Dextrins	Glucose
Maltase	Brush Border (Small Intestine)	Maltose Maltotriose	Glucose
Sucrase	Brush Border (Small Intestine)	Sucrose	Glucose Fructose
Lactase	Brush Border (Small Intestine)	Lactose	Glucose Galactose
PROTEIN – DIGESTING ENZYMES			
Pepsin	Chief Cells (Stomach)	Proteins	Peptides
Trypsin	Acinar Cells (Pancreas)	Proteins	Peptides
Chymotrypsin	Acinar Cells (Pancreas)	Proteins	Peptides
Carboxypeptidase	Acinar Cells (Pancreas)	Peptides (Carboxyl End)	Peptides Amino Acids
Aminopeptidase	Brush Border (Small Intestine)	Peptides (Amino End)	Peptides Amino Acids
Dipeptidase	Brush Border (Small Intestine)	Dipeptides	Amino Acids
LIPID – DIGESTING ENZYMES			
Lingual Lipase	Glands (Tongue)	Triglycerides	Fatty Acids Monoglycerides
Pancreatic Lipase	Acinar Cells (Pancreas)	Triglycerides	Fatty Acids Monoglycerides

DIGESTION AND ABSORPTION / Control Mechanisms

Salivation

Salivation is entirely under nervous control. Two nuclei in the brain stem (the superior and inferior *salivatory nuclei*) send parasympathetic impulses via the facial (VIII) and glossopharyngeal (IX) nerves to the salivary glands, stimulating the secretion of saliva. Chemicals in food stimulate taste buds on the tongue, which send nerve impulses to the brain stem and stimulate the salivatory nuclei. The salivatory nuclei can also be stimulated by the smell, sight, sound, or memory of food.

Swallowing (Deglutition)

During the second stage of swallowing (the pharyngeal stage), the passage of the bolus through the pharynx stimulates receptors in the oropharynx, which send impulses to the *deglutition center* in the medulla and lower pons of the brain stem. Impulses returning to the throat cause the changes that facilitate the passage of the bolus into the esophagus. During the third stage of swallowing (the esophageal stage) the bolus is pushed through the esophagus by peristalsis, which is controlled by the medulla.

Gastric Secretion and Motility

The secretion of gastric juice and the contraction of smooth muscle in the stomach are controlled by both nervous and hormonal mechanisms that occur in three overlapping phases:

(1) Cephalic Phase   The cephalic phase refers to the reflexes that are initiated by sight, smell, taste, chewing, and emotions. The cerebral cortex and *feeding center* in the hypothalamus send impulses to the medulla, which transmits impulses to the stomach via the vagus (X) nerves. The fibers of these parasympathetic nerves stimulate the secretion of gastric juice and promote gastric motility.

(2) Gastric Phase   The gastric phase refers to the reflexes that are initiated by stimuli in the stomach. Stretch receptors detect distension of the stomach and chemoreceptors detect an increase in the pH. Nerve impulses travel from the receptors to parasympathetic fibers in the *submucosal plexus*, causing waves of peristalsis and the secretion of gastric juice. Nerve impulses also stimulate secretion of gastrin, which stimulates the secretion of gastric juice and increased gastric emptying.

(3) Intestinal Phase   The intestinal phase refers to reflexes that are initiated by stimuli in the intestine. Distension of the duodenum and the presence of fatty acids and glucose stimulate *enteroendocrine cells* to secrete the hormones secretin, CCK, and GIP. These hormones inhibit secretion of gastric juice, gastric motility, and gastric emptying. The presence of food in the intestine also initiates a neural reflex called the *enterogastric reflex*. Nerve impulses carried to the medulla from the duodenum return to the stomach and inhibit gastric secretion and motility.

Intestinal Secretion and Motility

Local reflexes in response to the presence of chyme regulate secretion and motility in the small intestine. Distension of the intestinal wall initiates impulses to the *submucosal* and *myenteric plexuses* and the central nervous system. Local reflexes and returning parasympathetic impulses from the central nervous system (CNS) increase mixing of the chyme by segmentation. When most of the nutrients and water in a portion of the intestine have been absorbed, the decreased volume stimulates peristalsis, moving the undigested materials toward the large intestine.

Defecation

Mass peristaltic movements push fecal material from the sigmoid colon into the rectum. The resulting distension of the rectum initiates a reflex for defecation. Stretch receptors send nerve impulses to the *sacral spinal cord*. Returning parasympathetic impulses cause longitudinal rectal muscles to shorten (increasing the pressure in the rectum) and cause the internal anal sphincter to relax and open. Voluntary contractions of the diaphragm and abdominal muscles, and relaxation of the external anal sphincter results in defecation (feces are expelled through the anus).

HORMONAL CONTROL OF DIGESTION

Gastrin : secretion stimulated by peptides and amino acids in the stomach; stimulates the secretion of gastric juice (HCl and pepsinogen).

Secretin : secretion stimulated by acidic chyme in the duodenum; enhances the flow of bile rich in bicarbonate (HCO_3^-) from the liver.

CCK : secretion stimulated by amino acids and fatty acids in the duodenum; stimulates the secretion of bile and pancreatic digestive enzymes.

GIP : secretion stimulated by glucose and fat in the duodenum; inhibits the secretion of gastric juices.

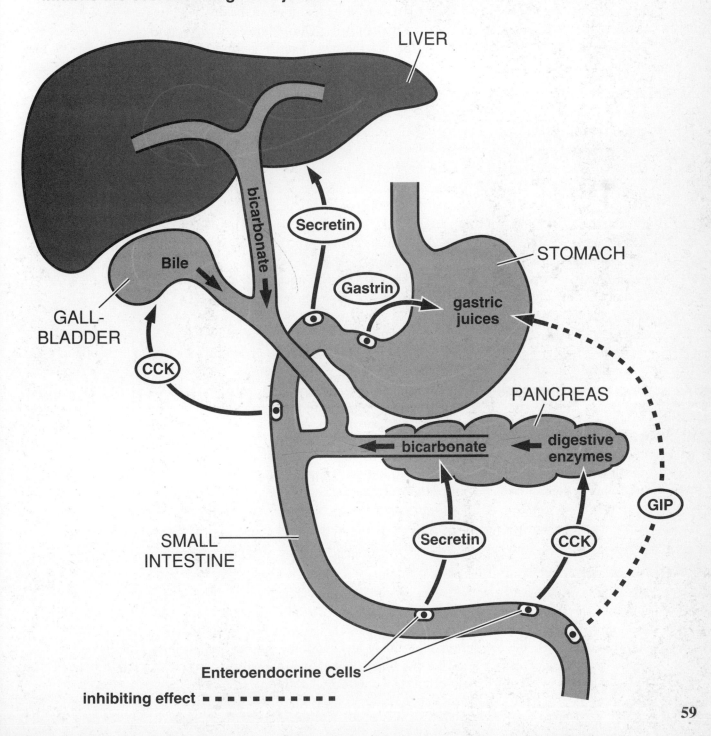

inhibiting effect ▪ ▪ ▪ ▪ ▪ ▪ ▪ ▪ ▪

59

4 Metabolism

METABOLISM / Overview

As blood passes through the capillaries of the tissues, cells take up nutrients by diffusion or active transport. Once inside the cells, they are either used as building blocks to make other molecules, stored, or broken down into carbon dioxide and water and used to produce ATP .

METABOLISM

The final products of digestion are: monosaccharides (simple sugars); fatty acids and monoglycerides (which are used to resynthesize triglycerides and are packaged in chylomicrons for transport in the blood); and amino acids. All of the chemical reactions that occur in body cells are collectively referred to as metabolism. There are two basic types of metabolism: anabolism (anabolic reactions) and catabolism (catabolic reactions).

Anabolism (Synthesis) Chemical reactions that combine simple organic molecules into more complex molecules are collectively known as anabolic reactions.

Catabolism (Breakdown) Chemical reactions that break down large organic molecules into simpler ones are collectively called catabolic reactions. When chemical bonds are split, energy is released; some of the energy released is transferred to and trapped in the high-energy phosphate bonds of ATP.

Coupling of Anabolism and Catabolism Catabolic reactions provide the energy (ATP) that most anabolic reactions require to proceed. The splitting of high-energy phosphate bonds in molecules of ATP releases energy in amounts that are usable for anabolic reactions.

ENERGY PRODUCTION
Phosphorylation

The addition of a phosphate group to a molecule is called phosphorylation; it raises the energy level of the molecule. In animal cells, ATP can be generated by substrate-level phosphorylation or oxidative phosphorylation.

Substrate Phosphorylation In substrate phosphorylation, ATP is generated when a high-energy phosphate group is transferred directly from a substrate (molecule involved in a series of reactions) to a molecule of ADP (adenosine diphosphate). Substrate phosphorylation produces 2 ATPs during glycolysis and 2 ATPs during the Krebs cycle.

Oxidative Phosphorylation In oxidative phosphorylation, electrons are removed from an organic molecule and passed along a series of molecules called electron acceptors to molecules of oxygen. The transfer of electrons from one electron acceptor to the next releases energy, which is used to attach a high-energy phosphate group to a molecule of ADP, forming ATP. Most of the ATP produced by cells is formed in this way. It occurs in the mitochondria.

Sources of ATP

Carbohydrates During the absorptive state (the first 4 hours after a meal) most body cells produce ATP by oxidizing glucose to carbon dioxide and water. The oxidation of glucose is called *cellular respiration*.

Lipids During the postabsorptive state (the period of time between meals when no glucose is being absorbed by the GI tract), most body cells use fatty acids as their main ATP source.

Proteins During prolonged fasting, large amounts of amino acids from protein breakdown are released from muscles and converted into glucose in the liver. The glucose is released into the general circulation and used for ATP production by body cells.

METABOLISM OVERVIEW

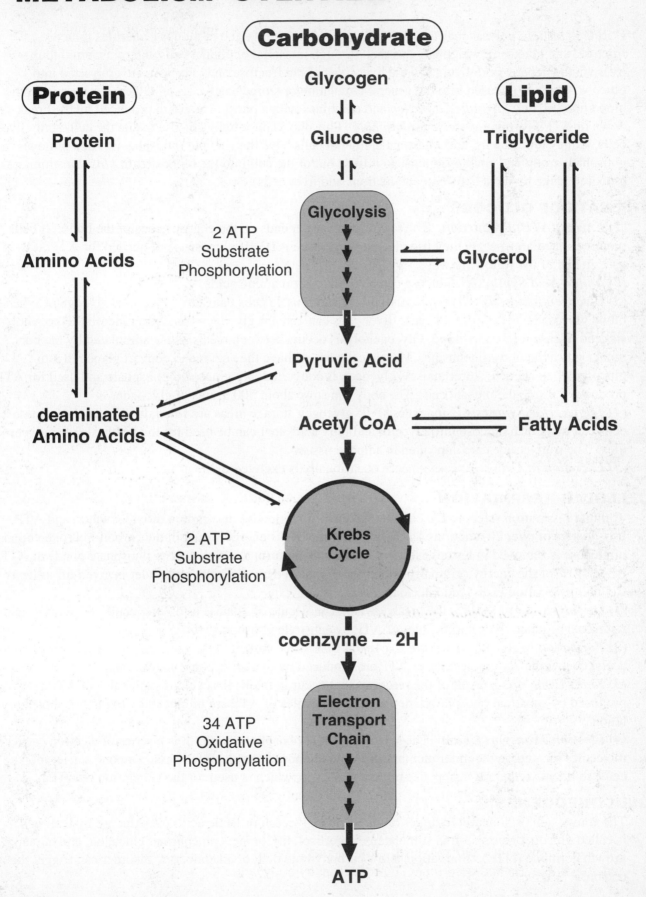

Carbohydrate

Protein

Lipid

Glycogen

Protein

Triglyceride

Glucose

Amino Acids

2 ATP
Substrate
Phosphorylation

Glycolysis

Glycerol

deaminated
Amino Acids

Pyruvic Acid

Acetyl CoA

Fatty Acids

2 ATP
Substrate
Phosphorylation

Krebs
Cycle

coenzyme — 2H

34 ATP
Oxidative
Phosphorylation

Electron
Transport
Chain

ATP

63

METABOLISM / Carbohydrates

During digestion, polysaccharides and disaccharides are converted to monosaccharides (glucose, fructose, and glactose), which are absorbed through intestinal epithelial cells and transported to the liver via the hepatic portal vein. In the liver most of the fructose and galactose are converted into glucose, which is released into the general circulation or stored as glycogen. Glucose moves from the blood into cells by *facilitated diffusion* and combines with a phosphate group produced by the break-down of ATP, forming *glucose 6-phosphate*. Phosphorylation traps glucose inside the cells; only liver cells, kidney tubule cells, and intestinal epithelial cells have the enzyme (phosphatase) that removes the phosphate group and enables glucose to diffuse out of the cell into the bloodstream. *Insulin* stimulates glucose uptake by most cells (except neurons and liver cells).

THE FATE OF GLUCOSE

(1) Energy (ATP) Production   The fate of glucose depends on the energy needs of the body. If cells require immediate energy, they use glucose to produce ATP. If glucose is not needed for ATP, it is used in one of the following ways.

(2) Amino Acid Synthesis   Glucose can be used to form amino acids.

(3) Glycogen Synthesis (Glycogenesis)   Liver and muscle cells can store glucose as glycogen. When blood glucose levels start to decrease, liver cells can convert glycogen back into glucose (glycogenolysis) and release it into the blood. Glycogenolysis occurs between meals and is stimulated by the hormones glucagon and epinephrine. Muscle cells do not have the enzyme to convert glycogen into glucose; when needed, stored muscle glycogen is converted into glucose 6-phosphate and used for ATP production inside the muscle cell. The body can store about 500 grams of glycogen.

(4) Triglyceride Synthesis (Lipogenesis)   If glycogen storage areas are filled, liver cells and adipose cells can transform glucose into glycerol and fatty acids that can be used for the synthesis of triglycerides. The triglycerides are deposited in adipose tissue.

(5) Excretion in Urine   Excess glucose occasionally is excreted in the urine.

CELLULAR RESPIRATION

Cellular respiration refers to the complete oxidation of glucose into carbon dioxide, water, and ATP. It is a series of over 20 reactions during which some of the energy stored in the carbon-hydrogen bonds of glucose is removed in a stepwise manner and used to form the high-energy phosphate bonds of ATP. About 40% of the energy originally in glucose is captured by ATP; the remainder is given off as heat. Cellular respiration is divided into four phases:

(1) Glycolysis   Glycolysis is the breakdown of 1 molecule of glucose into 2 molecules of pyruvic acid with a net production of 2 molecules of ATP by substrate phosphorylation.

(2) Formation of Acetyl Coenzyme A (Acetyl CoA)   As a result of the conversion of pyruvic acid into acetyl coenzyme A, 6 molecules of ATP are produced by oxidative phosphorylation.

(3) Krebs Cycle   As a result of the reactions that occur in the Krebs cycle, 2 molecules of ATP are produced by substrate phosphorylation and 22 molecules of ATP are produced by oxidative phosphorylation.

(4) Electron Transport Chain   The site of oxidative phosphorylation. It is a series of electron carrier molecules located on the inner membrane of mitochondria. As electrons pass from one carrier molecule to the next, there is a stepwise release of energy, which is used for the generation of ATP.

GLUCONEOGENESIS

The conversion of noncarbohydrates (amino acids, glycerol, or lactic acid) into glucose by liver cells is called gluconeogenesis. It is stimulated by cortisol, thyroxine, epinephrine, glucagon, and human growth hormone (hGH). Starvation, a low carbohydrate diet, or an endocrine disorder can trigger this response.

CARBOHYDRATE METABOLISM

Gluconeogenesis :
the synthesis of glucose from noncarbohydrate precursors

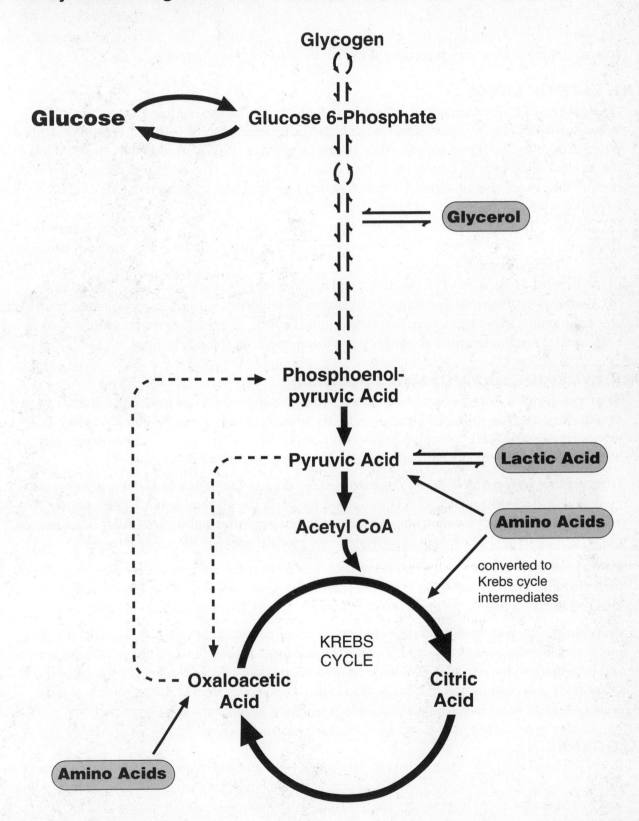

METABOLISM / Lipids

Within hours after eating a fatty meal, most chylomicrons (fat droplets) are removed from blood as they pass through capillaries in the liver and adipose tissue. An enzyme called *lipoprotein lipase* found in the endothelial cells that line these capillaries breaks down the triglycerides in chylomicrons into fatty acids and glycerol. The fatty acids diffuse into the liver and adipose cells and recombine with glycerol (produced by the cells) to form triglycerides.

THE FATE OF LIPIDS

(1) Energy (ATP) Production   Lipids, like carbohydrates, can be oxidized to produce ATP. When sufficient glucose is available in the blood (after meals), most cells preferentially oxidize glucose. During the postabsorptive state (when no glucose is being absorbed by the GI tract), most cells switch over to fatty acids as their main ATP source. The energy yield of triglycerides is more than twice that of carbohydrates: 1 gram of triglyceride produces 9 calories, while 1 gram of carbohydrate produces only 4 calories.

(2) Storage   The triglycerides stored in adipose tissue constitute 98% of all energy reserves. Excess dietary carbohydrates, proteins, and lipids all have the same fate: they are converted into triglycerides and stored in adipose tissues.

(3) Structural Molecules   Phospholipids and cholesterol are components of cell membranes.

(4) Synthesis of Essential Substances   Examples of important substances synthesized from lipids are: lipoproteins (lipid transport), bile salts (emulsification of lipids), steroid hormones (sex hormones and adrenocortical hormones), and thromboplastin (blood clotting).

TRIGLYCERIDE BREAKDOWN (Lipolysis)

Triglycerides into Glycerol and Fatty Acids   Before triglycerides can be used as an energy source, they must be split into glycerol and fatty acids. During times of glucose scarcity, hormones (epinephrine, norepinephrine, glucocorticoids, thyroid hormones, and human growth hormone) stimulate the breakdown of triglycerides into glycerol and fatty acids.

Glycerol into Glyceraldehyde 3-Phosphate   Many cells of the body can convert glycerol into glyceraldehyde 3-phosphate. If ATP is needed by a cell, the glyceraldehye 3-phosphate is converted into pyruvic acid, which is part of the cellular respiration pathway. If ATP is not needed, the glyceraldehyde 3-phosphate is converted into glucose and then glycogen (storage).

Fatty Acids into Acetyl CoA (Beta Oxidation)   In beta oxidation, enzymes remove one pair of carbon atoms at a time from the fatty acid chains; the resulting two-carbon fragment then is attached to coenzyme A, forming acetyl coenzyme A, which enters the Krebs cycle.

Fatty Acids into Ketone Bodies (Ketogenesis)   Liver cells can use two acetyl coenzyme A molecules to form acetoacetic acid. Some acetoacetic acid molecules are converted into acetone and beta-hydroxybutyric acid. All three substances are collectively known as ketone bodies. Other body cells can use acetoacetic acid to make acetyl coenzyme A. Heart muscle and the cortex of the kidneys use acetoacetic acid in preference to glucose for ATP production.

LIPOGENESIS

Liver cells and adipose tissue can synthesize lipids from glucose or amino acids through a process called lipogenesis. This occurs when you consume more calories than are needed to satisfy your energy requirements. The resulting triglycerides can be stored or used to produce other lipids such as cholesterol, lipoproteins, or phospholipids.

LIPID METABOLISM

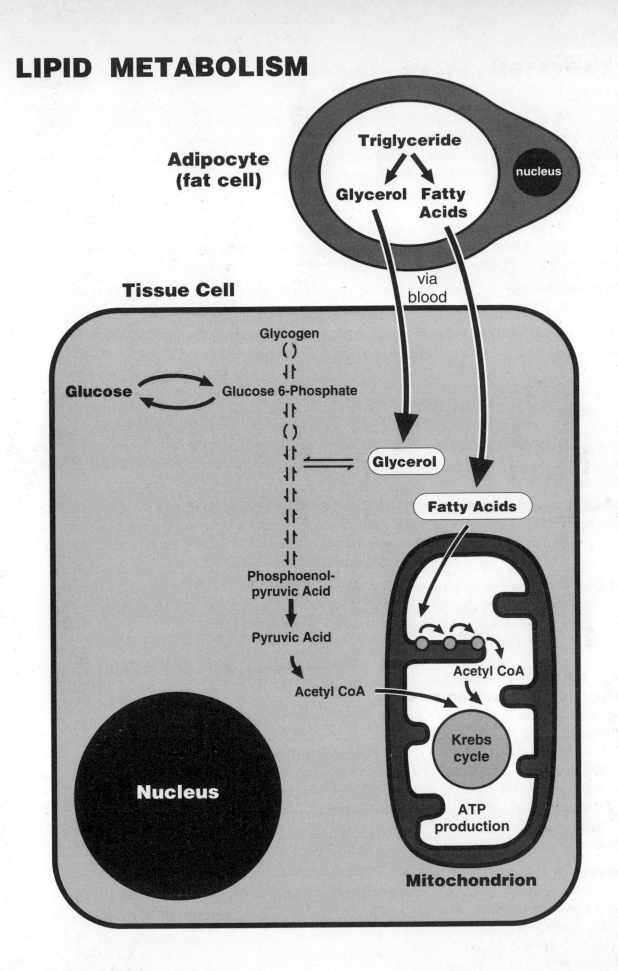

METABOLISM / Plasma Lipids

High blood cholesterol promotes the growth of fatty plaques that build up in the walls of arteries. As a plaque enlarges, the passageway for blood progressively narrows, which restricts blood flow; this causes a backup of blood, which increases arterial blood pressure. The rough surface of the plaque also promotes blood clotting, which can cause a heart attack or stroke.

LIPOPROTEINS

When cholesterol and triglycerides are combined with proteins produced by the liver, they become water-soluble. These combinations of lipids and proteins (lipoproteins) make it possible for lipids to be transported in blood plasma (the water portion of blood).

Chylomicrons Chylomicrons (protein-coated fat droplets) carry lipids from the epithelial cells that line the GI tract via lymphatic vessels and blood vessels to the tissue cells of the body.

LDLs ("bad" cholesterol) High levels of low-density lipoproteins (LDLs) are associated with increased risk of atherosclerosis. LDLs deliver cholesterol to body cells that need it and deposit cholesterol in and around smooth muscle fibers in arteries, forming fatty plaques. LDLs attach to receptors on the surfaces of cells and are taken into the cells by receptor-mediated endocytosis. Cholesterol is released inside the cells. Once a cell has sufficient cholesterol, a negative feedback system inhibits the cell from synthesizing new LDL receptors. As a result, there is an increase in the plasma LDL level, making it more likely that fatty plaques will develop in the arteries. Environmental and genetic factors can influence the number of LDL receptors on an individual's cells; and this affects their plasma LDL levels.

HDLs ("good" cholesterol) High levels of high-density lipoproteins (HDLs) are associated with decreased risk of atherosclerosis. HDLs remove excess cholesterol from body cells and transport it to the liver for elimination. This pick-up service prevents accumulation of cholesterol in the blood, reducing the risk of fatty plaque formation in the arteries.

VLDLs Very low-density lipoproteins (VLDLs) transport triglycerides synthesized by liver cells to adipose cells for storage. A high fat diet promotes the production of VLDLs. After depositing their triglycerides in adipose cells, the VLDLs are converted to LDLs. In this way a high fat diet may lead to an increased risk of fatty plaque formation in arteries.

BLOOD CHOLESTEROL
Sources of Blood Cholesterol

There are two sources of blood cholesterol: foods and liver synthesis.

(1) Foods (eggs, dairy products, organ meats, beef, pork, and processed luncheon meats)

Fatty foods that do not contain any cholesterol can still increase blood cholesterol level in two ways:

Dietary Fats A high intake of dietary fats stimulates the reabsorption of *cholesterol-containing bile* back into the blood; consequently, less cholesterol is lost in the feces.

Saturated Fats When saturated fats are broken down, the liver uses some of the breakdown products to produce cholesterol.

(2) Liver Synthesis Most of the cholesterol in the body is synthesized by the liver.

Levels of Blood Cholesterol

For adults, desirable levels of blood cholesterol (measured in milligrams per deciliter) are:

Total Cholesterol (TC) : under 200 mg/dl

LDL – Cholesterol : under 130 mg/dl

HDL – Cholesterol : over 40 mg/dl

Triglycerides : 10 – 190 mg/dl

ATHEROSCLEROSIS

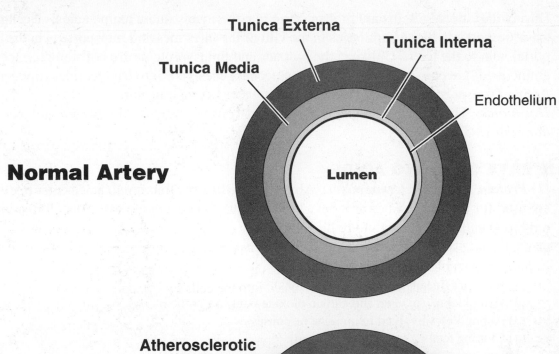

Tunica Externa

Tunica Interna

Tunica Media

Endothelium

Normal Artery

Lumen

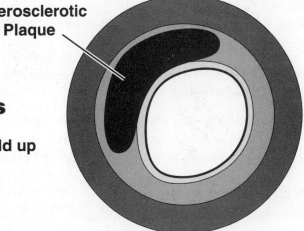

Atherosclerotic
Plaque

Moderate
Atherosclerosis

**Fatty deposits begin to build up
in the tunica media**

Extreme
Atherosclerosis

**Fatty deposits and calcium buildup;
blockage almost total**

METABOLISM / Proteins

During digestion, proteins are broken down into their constituent amino acids. The amino acids are absorbed into blood capillaries in the villi of the intestines and transported via the hepatic portal vein to the liver. Unlike carbohydrates and lipids, proteins are not stored. Excess dietary amino acids are converted into glucose (gluconeogenesis) or into triglycerides (lipogenesis).

Active Transport Amino acids enter body cells by active transport.

Hormones The uptake of amino acids by body cells is stimulated by *insulin* and *human growth hormone (hGH)*.

THE FATE OF AMINO ACIDS

(1) Protein Synthesis (Anabolism) Almost immediately after amino acids enter body cells they are used for the synthesis of proteins. Protein synthesis occurs in most cells. It does not occur in red blood cells, since a mature red blood cell has no nucleus (the DNA in a nucleus controls protein synthesis). Many different substances essential to cells are made of protein. Some important examples are the following:

    (1) Enzymes (catalyze chemical reactions).
    (2) Hemoglobin (oxygen and carbon dioxide transport in the blood).
    (3) Antibodies (involved in immune responses).
    (4) Clotting factors (such as fibrinogen).
    (5) Hormones (all hormones except sex hormones and mineralocorticoids).
    (6) Neurotransmitters.
    (7) Contractile elements in muscle fibers (actin and myosin).
    (8) Structural components (collagen, elastin, keratin, and nucleoproteins).
    (9) Plasma membrane receptors.

(2) Energy (ATP) Production If other energy sources are used up or if other sources are inadequate for the needs of the body, the liver can convert proteins into fatty acids, ketone bodies, or glucose. These substances can be used by cells for the production of ATP.

AMINO ACID CONVERSIONS

Amino acids can be converted into keto acids, new amino acids, glucose, or fatty acids.

Amino Acids into Keto Acids (Deamination) During the process called oxidative deamination an amino group ($-NH_2$) is removed from an amino acid, producing a keto acid (a type of carbohydrate). The keto acid can be used for ATP production or the synthesis of fatty acids. The amino group removed during deamination is converted into ammonia (NH_3). The ammonia is carried by the blood to the liver, where it is converted into urea. Urea is carried by the blood to the kidneys, where it is excreted in the urine.

Amino Acids into New Amino Acids (Transamination) Transamination is the transfer of an amino group ($-NH_2$) from an amino acid to a keto acid. With the added amino group, the keto acid now becomes an amino acid.

Amino Acids into Glucose (Gluconeogenesis) Liver cells can convert amino acids into glucose.

Amino Acids into Fatty Acids (Lipogenesis) Liver cells and adipose cells can convert amino acids into fatty acids.

PROTEIN METABOLISM

Deamination

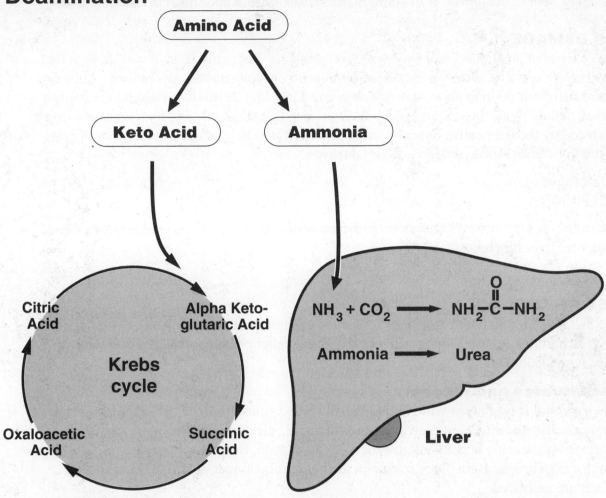

Example : Deamination of Glutamic Acid

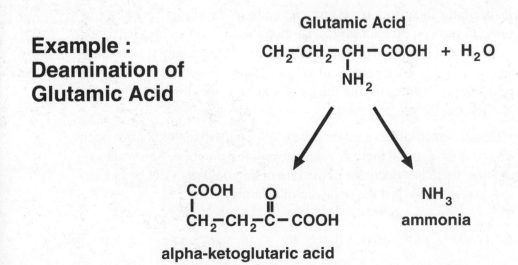

Glutamic Acid

$$CH_2\!-\!CH_2\!-\!CH\!-\!COOH \ + \ H_2O$$
$$\underset{NH_2}{\big|}$$

alpha-ketoglutaric acid

$$\underset{CH_2\!-\!CH_2\!-\!C\!-\!COOH}{\overset{COOH \quad\quad O}{\big| \qquad\qquad \|}}$$

NH_3
ammonia

METABOLISM / Alcohol Biotransformation

In the United States, alcoholism is the third major cause of death after heart disease and cancer.

LIVER DAMAGE
The overuse of alcohol is associated with liver damage. For many years it was thought that this damage was due to the malnutrition so frequently accompanying alcoholism. Although severe malnutrition may play some role, it is now clear that the toxic damage to the liver is caused mainly by the products of alcohol catabolism (acetaldehyde and hydrogen), which are produced by the liver cells. Other organs and tissues are also damaged in a variety of ways by the toxic effects of hydrogen and acetaldehyde.

Acetaldehyde Alcohol is initially broken down by liver cells to hydrogen and acetaldehyde. Acetaldehyde damages mitochondria, effecting the production of ATP.

Hydrogen The excess hydrogen causes the accumulation of fat in liver cells, which interferes with liver functions.

ABUSE OF ALCOHOL
The metabolism of alcohol also leads to marked changes in the metabolism of other drugs (such as barbiturates) and foreign chemicals. These other chemicals share metabolic pathways with alcohol.

Barbiturates and Alcohol
When alcohol is first taken with barbiturate, the blood concentrations of both drugs increase; consequently, the effects of both drugs are enhanced. Because the alcohol and barbiturate compete for the same hepatic microsomal enzyme system, there is a decreased rate of catabolism for both drugs. Both drugs remain in the blood for a longer period of time, so their effects are enhanced.

Chronic Overuse of Alcohol
Decreased Effects ("Tolerance") Chronic overuse of alcohol (before significant liver damage has occurred) stimulates the microsomal enzyme system; the chronic presence of alcohol causes an increase in the system's activity. The increased rate of catabolism due to increased enzyme activity lowers the blood concentration of alcohol, making it necessary to drink more in order to achieve a given magnitude of effect. Many alcoholics can tolerate more alcohol than their drinking partners during the early stages of the disease. This situation reverses itself as the liver cells are destroyed.

Increased Effects (increased sensitivity to a given dose) When the liver cells have been severely damaged by overuse of alcohol, there is a decrease in the number of enzyme systems still functioning. Thus, there is a decrease in the rate of catabolism of drugs and an increased sensitivity to a given dose—just the opposite of tolerance.

ALCOHOL BIOTRANSFORMATION

Breakdown of Alcohol by Liver Cells

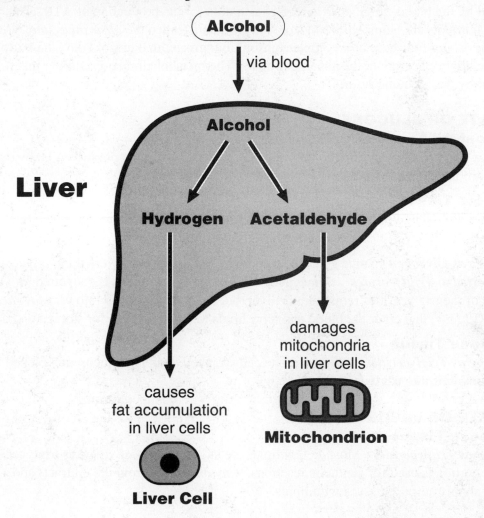

Chronic Overuse of Alcohol

Stage 1 : Tolerance

Before Liver Damage

chronic overuse
stimulates
microsomal enzyme systems

⬇

increased
rate of catabolism
of alcohol

Stage 2 : Increased Sensitivity

After Liver Damage

decreased number
of functioning
microsomal enzyme systems

⬇

decreased
rate of catabolism
of alcohol

METABOLISM / Absorptive (Fed) State

After a meal (for about four hours) ingested nutrients are present in the GI tract and are being absorbed by the blood, so glucose is readily available for the production of ATP. This period of time when nutrients are being absorbed from the GI tract is called the *absorptive state*. An average meal requires about four hours for complete absorption; given three meals a day, the body spends about twelve hours each day in the absorptive state. The metabolic reactions during this period are controlled by the hormone *insulin*.

THE FATE OF GLUCOSE
Almost All Tissues
Energy Production   During the absorptive state most body cells produce ATP by oxidizing glucose into carbon dioxide and water.

Muscle Tissue
Storage as Glycogen   Muscles cells take up some glucose and store it as glycogen.

Liver
Storage as Glycogen   Some glucose taken up by liver cells is converted into glycogen.
Converted to Triglycerides   Some glucose taken up by liver cells is converted into triglycerides. Some of the triglycerides remain in the liver, but most are packaged into *very low-denstiy lipoproteins (VLDLs)* that enter the blood and ferry lipids to adipose tissue for storage.

Adipose Tissue
Storage as Triglycerides   Adipose tissue cells (fat cells) take up glucose not picked up by the liver and convert it into triglycerides for storage.

THE FATE OF LIPIDS
Adipose Tissue
Stored as Triglycerides   Most dietary lipids are stored in adipose tissue as triglycerides. Only a small portion is used for synthetic reactions. *Chylomicrons* (from the GI tract) and *VLDLs* (from the liver) supply adipose tissues with lipids.

THE FATE OF AMINO ACIDS
Almost All Tissues
Protein Synthesis   Some amino acids enter body cells, such as muscle cells for synthesis of proteins, such as enzymes and hormones.

Liver
Deamination   Many amino acids that enter liver cells are deaminated (amino groups removed), forming *keto acids* (a type of carbohydrate). The keto acids can enter the Krebs cycle for ATP production or be used to synthesize glucose or fatty acids.
Protein Synthesis   Some amino acids that enter liver cells are used to synthesize proteins, such as *plasma proteins*.

HORMONAL REGULATION
Effects of Insulin
(1) Promotes the entry of glucose and amino acids into the cells of many tissues.
(2) Stimulates phosphorylation of glucose in liver cells.
(3) Stimulates conversion of glucose 6-phosphate to glycogen in liver and muscle cells.
(4) Enhances the synthesis of triglycerides in liver and adipose tissue.

ABSORPTIVE (FED) STATE

The Fate of Glucose

Almost All Tissues :
Energy (ATP) Production

Glucose $\longrightarrow$ $CO_2 + H_2O + ATP$

Muscle Tissue :
Stored as Glycogen

Glucose $\longrightarrow$ Glycogen

Liver :
Stored as Glycogen
Converted to Triglycerides

Glucose $\longrightarrow$ Glycogen

Glucose $\longrightarrow$ Triglycerides

Adipose Tissue :
Stored as Triglycerides

Glucose $\longrightarrow$ Triglycerides

VLDLs

The Fate of Triglycerides

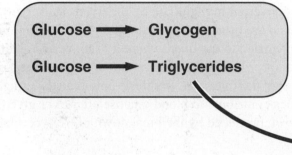

VLDLs
from the liver

Chylomicrons
from the GI tract

lipoprotein
lipase

Fatty
Acids

Adipose Tissue :
Stored as Triglycerides

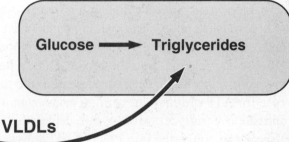

Fatty Acids
+
Glyceraldehyde
3-phosphate

$\longrightarrow$ Triglycerides

The Fate of Amino Acids

Almost All Tissues :
Protein Synthesis

Amino Acids $\longrightarrow$ Protein

Liver :
Deamination
Protein Synthesis

Amino Acids $\longrightarrow$ Keto Acids

Amino Acids $\longrightarrow$ Protein

METABOLISM / Postabsorptive (Fasting) State

Between meals (starting about four hours after a meal) nutrients are not present in the GI tract, so energy needs must be satisfied by substances stored in the cells of the body, primarily triglycerides and glycogen. This period of time is called the *postabsorptive state* ("after" absorption). Since cells are constantly removing glucose from the blood and no glucose is entering the blood from the GI tract, the principal metabolic challenge is to maintain a normal blood glucose level of 70 – 110 mg/100 ml of blood. This is especially important to the nervous system, since neurons cannot use other nutrients for ATP production (except ketone bodies during starvation). Blood glucose concentration is maintained in two ways: by synthesizing glucose and by glucose-sparing reactions (reactions that use lipids for ATP production).

GLUCOSE SYNTHESIS
Liver
Glycogen Breakdown (Glycogenolysis) A major source of blood glucose during fasting is provided by the breakdown of liver glycogen; it can provide a four hour supply of glucose.

Glucose Synthesis from Noncarbohydrates (Gluconeogenesis) Noncarbohydrates, such as pyruvic acid, lactic acid, amino acids, and glycerol, can be converted into glucose by the liver. During periods of vigorous exercise significant amounts of *lactic acid* are produced by anaerobic glycolysis in muscle cells. This lactic acid is released into the blood and carried to the liver, where it is converted again into glucose and released into the blood. During periods of prolonged fasting, *amino acids* (from protein breakdown in muscle cells) are released into the blood; the liver converts these amino acids into glucose for ATP production. Amino acids from muscles contribute to blood glucose after liver glycogen and adipose triglyceride stores are depleted. *Glycerol*, produced by the breakdown of triglycerides, also can be converted into glucose by the liver.

GLUCOSE–SPARING REACTIONS
During the postabsorptive state all body cells (except neurons) reduce their oxidation of glucose and switch over to fatty acids as their main source of ATP. This "spares" the available glucose for utilization by neurons.

Adipose Tissue
Triglyceride Breakdown (Lipolysis) Triglycerides in adipose tissue are broken down and fatty acids are released into the blood. These fatty acids are taken up by most body cells, converted into acetyl coenzyme A, and entered into the Krebs cycle for ATP production. (The glycerol released by lipolysis is converted into glucose by liver cells.)

Liver
Fatty Acids into Ketone Bodies The liver takes up some of the fatty acids released from adipose tissue and converts them into ketone bodies.

Almost All Tissues
Oxidation of Fatty Acids and Ketone Bodies The oxidation of fatty acids and the ketone bodies produced from them by the liver provides most of the ATP for body cells.

HORMONAL REGULATION
Glycogen Breakdown (Glycogenolysis) Stimulated by glucagon and epinephrine.

Glucose Synthesis from Noncarbohydrates (Gluconeogenesis) Stimulated by cortisol and glucagon.

Triglyceride Breakdown (Lipolysis) Stimulated by epinephrine, norepinephrine, cortisol, human growth hormone, and thyroid hormones.

Protein Breakdown Stimulated by cortisol.

POSTABSORPTIVE (FASTING) STATE
Maintenance of the Normal Blood Glucose Level (70 to 110 mg/ml of Blood)

Glucose Synthesis

Neurons require glucose for ATP production (they also can use ketone bodies during starvation). During the postabsorptive state glucose is synthesized by the liver from liver glycogen and several noncarbohydrate precursors.

Liver :
Glycogenolysis
(glycogen breakdown)

Gluconeogenesis
(glucose synthesis from noncarbohydrates)

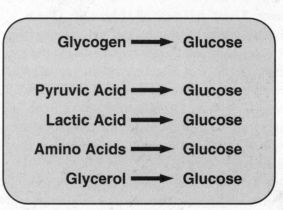

Glycogen ⟶ Glucose

Pyruvic Acid ⟶ Glucose

Lactic Acid ⟶ Glucose

Amino Acids ⟶ Glucose

Glycerol ⟶ Glucose

Glucose-Sparing Reactions

During the postabsorptive state all body cells (except neurons) reduce their oxidation of glucose and switch over to fatty acids as their main source of ATP. This "spares" the available glucose for utilization by neurons.

Adipose Tissue :
Lipolysis
(triglyceride breakdown)

Triglycerides ⟶ Fatty Acids + Glycerol

Almost All Tissues (except nervous) :
Energy (ATP) Production

Fatty Acids ⟶ $CO_2 + H_2O + ATP$

Ketone Bodies ⟶ $CO_2 + H_2O + ATP$

Liver :
Conversions

Fatty Acids ⟶ Ketone Bodies

METABOLISM / Body Heat

Heat   Heat is a form of kinetic energy. It can be measured as *temperature* and is expressed in units called *calories*.

Calorie (cal)   A calorie is the amount of heat energy required to raise the temperature of 1 gram of water 1 degree centigrade.

Kilocalorie (kcal)   Since the calorie is a small unit relative to the amount of energy stored in foods, the kilocalorie is used instead. A kilocalorie is equal to 1,000 calories. Although they are called calories, dietary calories are actually kilocalories.

METABOLIC RATE
Factors That Affect Metabolic Rate

Most of the heat generated by the body comes from the metabolism (oxidation) of foods. The rate at which this heat is generated is called the *metabolic rate*, and is measured in kilocalories. The major factors that affect metabolic rate are the following:

(1) Exercise   During strenuous exercise, the metabolic rate can increase 15-fold.

(2) Nervous System   During stress, nerve fibers of the sympathetic nervous system release norepinephrine, which increases the metabolic rate.

(3) Hormones   Epinephrine, thyroid hormones, testosterone, and human growth hormone increase metabolic rate.

(4) Body Temperature   The higher the body temperature, the higher the metabolic rate.

(5) Ingestion of Food   Ingestion of food (especially protein) increases the metabolic rate.

(6) Age   The metabolic rate of a child is double that of an elderly person.

Basal Metabolic Rate (BMR)

The basal metabolic rate (BMR) is a measure of the rate at which the quiet, resting, fasting body breaks down nutrients to liberate energy. It is most often measured indirectly by measuring oxygen consumption using a spirometer.

Thyroid Hormones   Thyroid hormones are the main regulator of the basal rate of ATP use. When the thyroid gland is secreting large quantities of thyroid hormones, the BMR can double; when it is secreting small quantities of thyroid hormones, the BMR can be halved.

REGULATION OF BODY TEMPERATURE

Core Temperature   The body's temperature below the skin and subcutaneous tissue. The normal core temperature is about 37° C (98.6° F).

Shell Temperature   The body's temperature at the surface (the skin and subcutaneous tissue).

Preoptic Area   A group of neurons in the hypothalamus that control temperature reflexes.

Heat-Losing Center (in hypothalamus)   Controls responses that lower body temperature.

Heat-Promoting Center (in hypothalamus)   Controls responses that raise body temperature.

Heat Loss

If the heat produced during the oxidation of foods is not removed, the body temperature will rise. The principal routes of heat loss are radiation, conduction, convection, and evaporation.

Heat Production

If the body temperature falls below normal, thermoreceptors in the skin and hypothalamus send nerve impulses to control centers (preoptic area and heat-promoting center) in the hypothalamus. Sympathetic nerve impulses from the heat-promoting center causes blood vessels to constrict (decreases heat loss through the skin), skeletal muscles to contract in a repetitive cycle called shivering, and the adrenal medulla to secrete epinephrine (increases cellular metabolism). Thyroid-releasing hormone (secreted by the anterior pituitary) stimulates the thyroid gland to secrete thyroid hormones, which increase the metabolic rate.

HORMONES THAT AFFECT METABOLISM

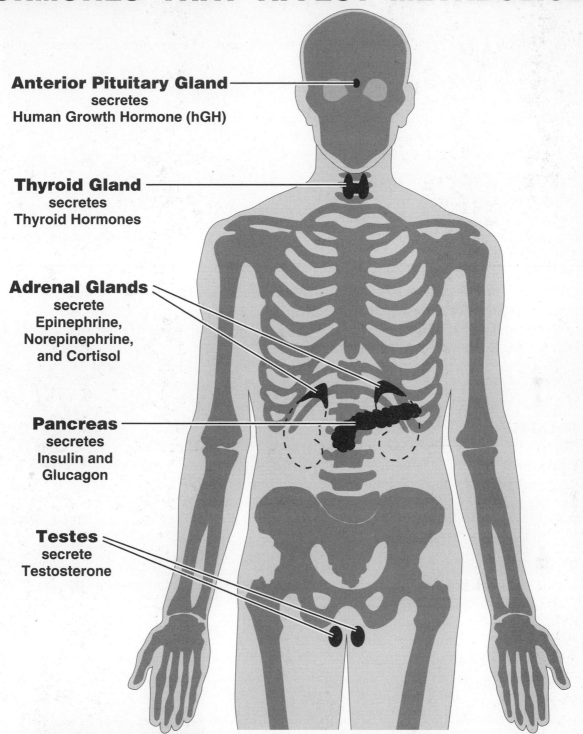

Anterior Pituitary Gland
secretes
Human Growth Hormone (hGH)

Thyroid Gland
secretes
Thyroid Hormones

Adrenal Glands
secrete
Epinephrine,
Norepinephrine,
and Cortisol

Pancreas
secretes
Insulin and
Glucagon

Testes
secrete
Testosterone

Metabolic Functions:
Human Growth Hormone : glycogenesis; lipolysis; protein synthesis.
Thyroid Hormones : lipolysis; protein synthesis; increase metabolic rate.
Epinephrine : glycogenolysis; lipolysis; increases metabolic rate.
Norepinephrine : lipolysis.
Cortisol : gluconeogenesis; lipolysis; protein synthesis.
Insulin : glycogenesis; lipogenesis; protein synthesis.
Glucagon : gluconeogenesis.
Testosterone : increases metabolic rate.

79

Part II : Self-Testing Exercises

Unlabeled illustrations from Part I

NUTRITIONAL GUIDELINES

CALORIC VALUE OF BASIC NUTRIENTS

Lipids (Fats) : ___ Calories per gram
Carbohydrates : ___ Calories per gram
Proteins : ___ Calories per gram

THE BASIC DIET

Recommended by the American Heart Association
for the average person who wants to maintain
normal blood cholesterol and lipid levels.

Percent of Total Calories

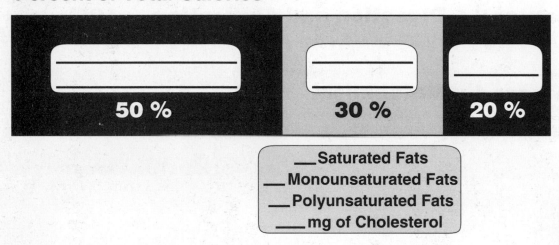

50 %   30 %   20 %

___Saturated Fats
___Monounsaturated Fats
___Polyunsaturated Fats
___mg of Cholesterol

DIETARY GUIDELINES

Recommended by the Committee on Diet and Health of the
Food and Nutrition Board (under the National Research Council).

(1) Total Fat Intake : reduce to ___percent or less of total calories.

(2) Complex Carbohydrates : eat ___or more daily servings.

(3) Fruits and Vegetables : eat ___or more daily servings.

(4) Proteins : maintain protein consumption at _____ levels.

(5) Exercise : balance food intake with exercise to maintain _____.

(6) Alcohol : do not drink alcohol; or limit the amount to ___drinks daily.

(7) Salt : limit the amount to ___grams (slightly less than 1 teaspoon).

(8) Calcium : maintain an adequate calcium intake.

(9) Dietary Supplements : Avoid _____of U.S. RDAs in any one day.

(10) Fluoride : maintain an_____level of fluoride in the diet.

CARBOHYDRATES

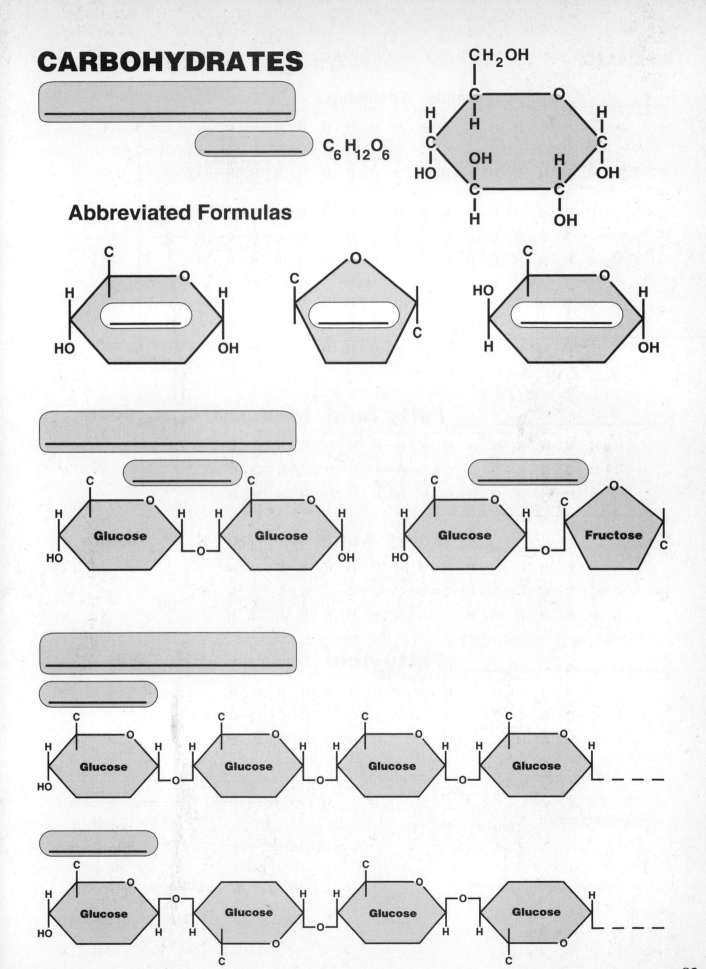

$$C_6 H_{12} O_6$$

Abbreviated Formulas

Glucose

Glucose   Glucose

Glucose   Fructose

Glucose   Glucose   Glucose   Glucose

Glucose   Glucose   Glucose   Glucose

LIPIDS

_____ (example : Tristearin)

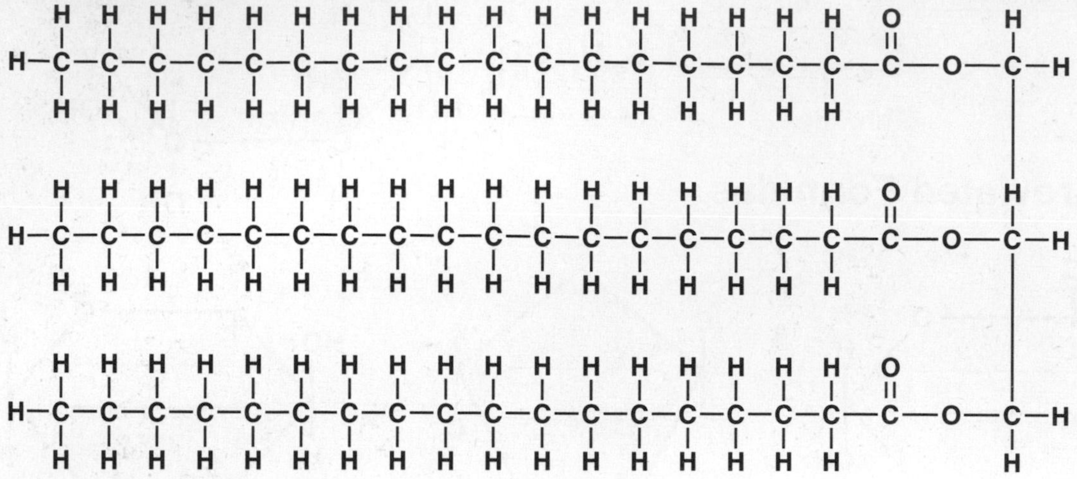

Fatty Acid Stearic Acid $C_{17}H_{35}COOH$

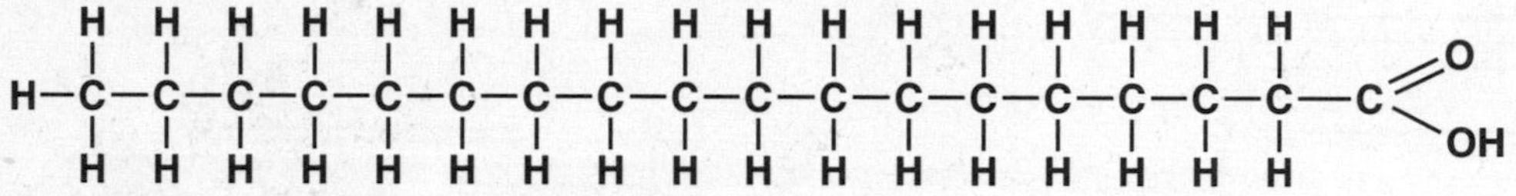

Fatty Acid Oleic Acid $C_{17}H_{33}COOH$

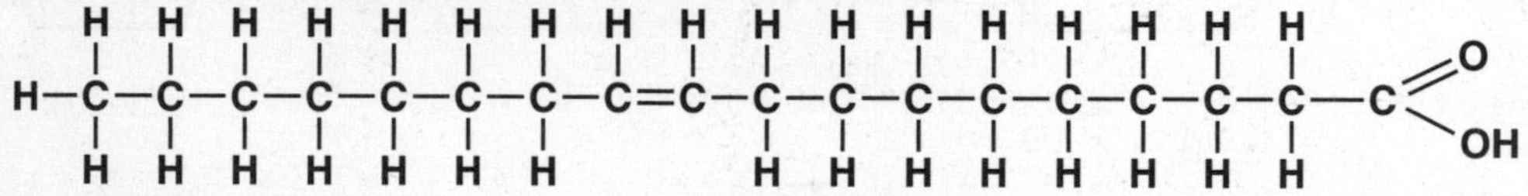

Fatty Acid Trilinolein $C_{17}H_{31}COOH$

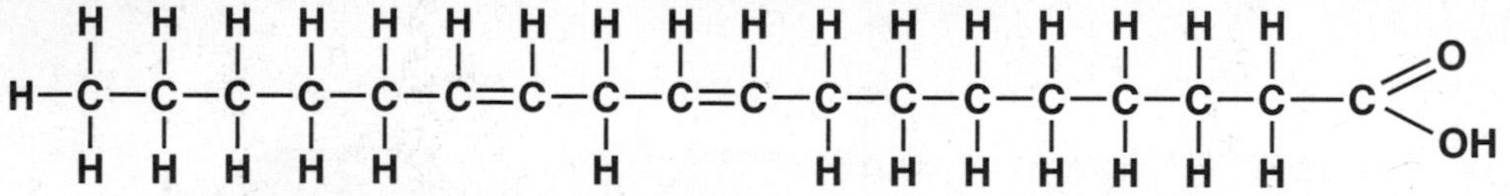

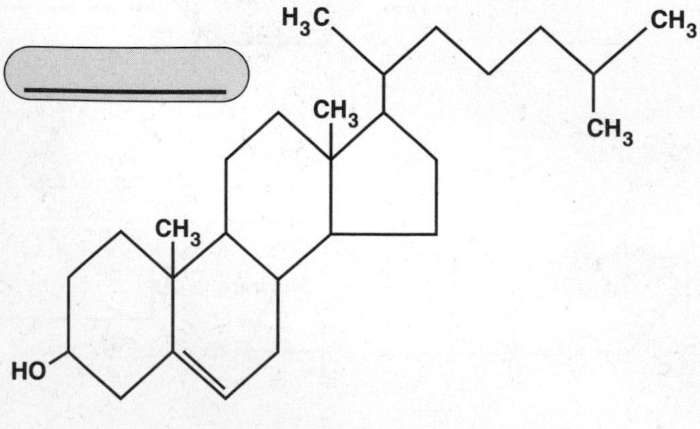

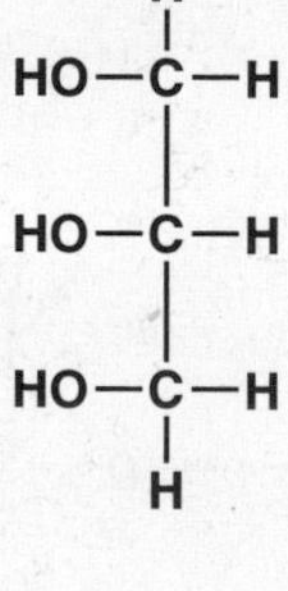

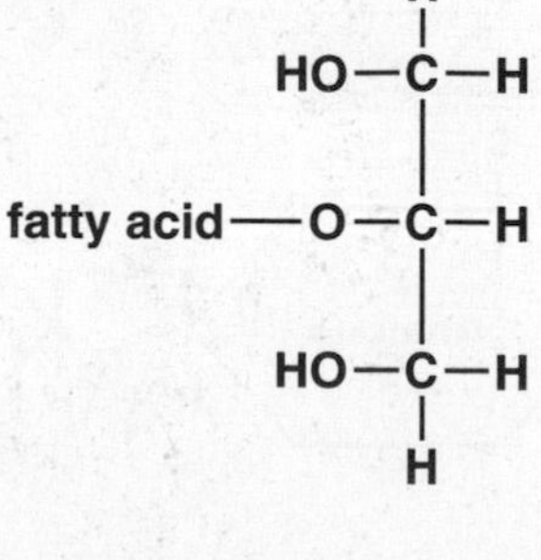

84

PROTEINS

| | 20 different _____ are found in proteins; 8 _____ are essential (must be in dietary foods). |

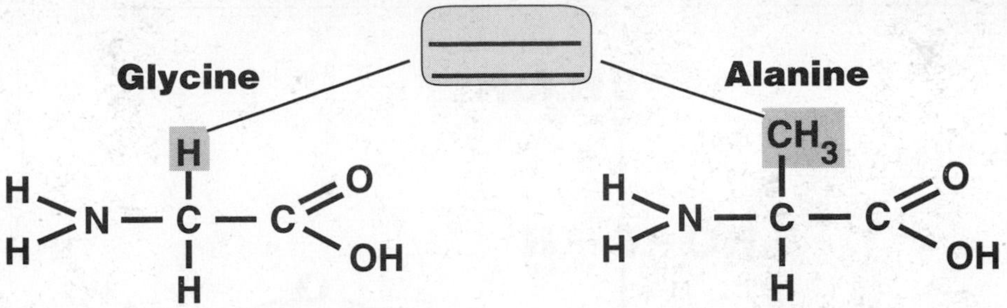

_____

Glycine

```
    H
    |
H   |      O
 \  |     //
  N-C----C
 /  |     \
H   |      OH
    H
```

Alanine

```
        CH₃
         |
H        |      O
 \       |     //
  N------C----C
 /       |     \
H        |      OH
         H
```

_____

2 amino acids linked by a _____ bond.

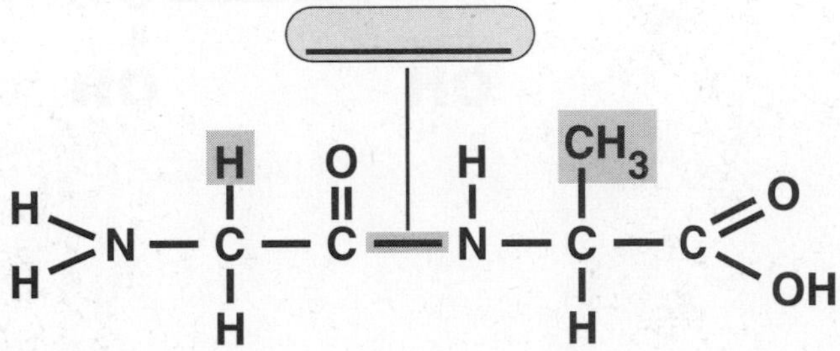

_____

```
    H      O        H    CH₃
    |      ||       |     |
H   |      ||       |     |      O
 \  |      ||       |     |     //
  N-C------C====N---C----C
 /  |              |      \
H   |              |       OH
    H              H
```

_____

A _____ consists of fewer than 50 amino acids.

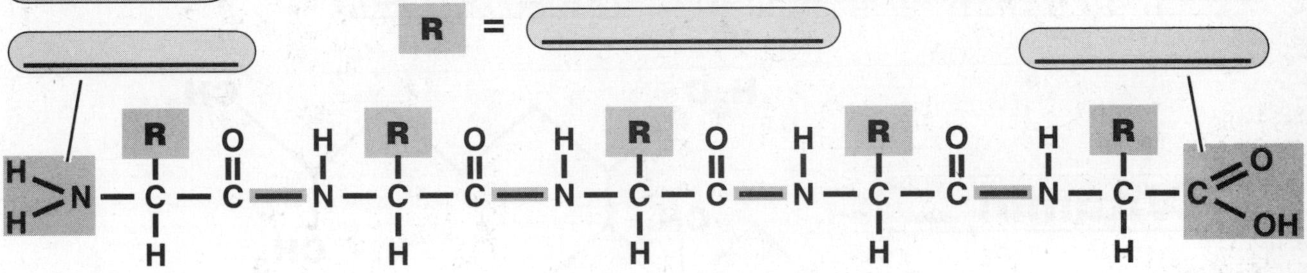

R = _____

_____   _____

```
    R   O   H   R   O   H   R   O   H   R   O   H   R
    |   ||  |   |   ||  |   |   ||  |   |   ||  |   |      O
H   |   ||  |   |   ||  |   |   ||  |   |   ||  |   |     //
 \  |   ||  |   |   ||  |   |   ||  |   |   ||  |   |    C
  N-C---C===N---C---C===N---C---C===N---C---C===N---C---  \
 /  |       |   |       |   |       |   |       |   |       OH
H   |       |   |       |   |       |   |       |   |
    H       H   H       H   H       H   H       H   H
```

_____

A _____ consists of more than 50 amino acids.

● = _____

_____

VITAMINS

_____-soluble Vitamins

_____-soluble vitamins include the B vitamins and vitamin __.

Vitamin __

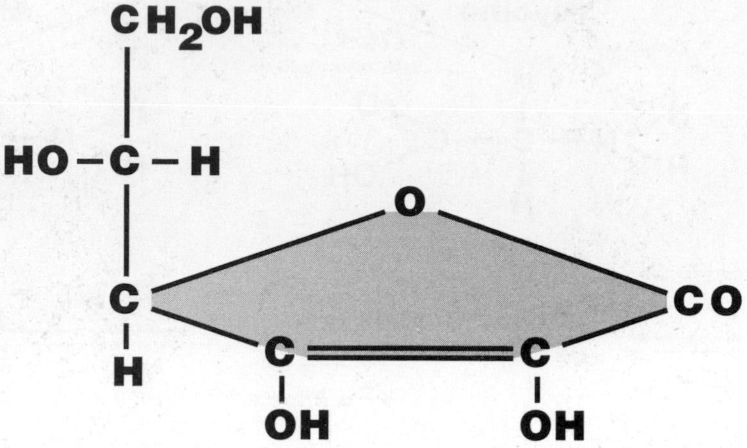

$$CH_2OH$$

$$HO-C-H$$

_____-soluble Vitamins

_____-soluble vitamins include vitamins A, D, __, and __.

Vitamin __

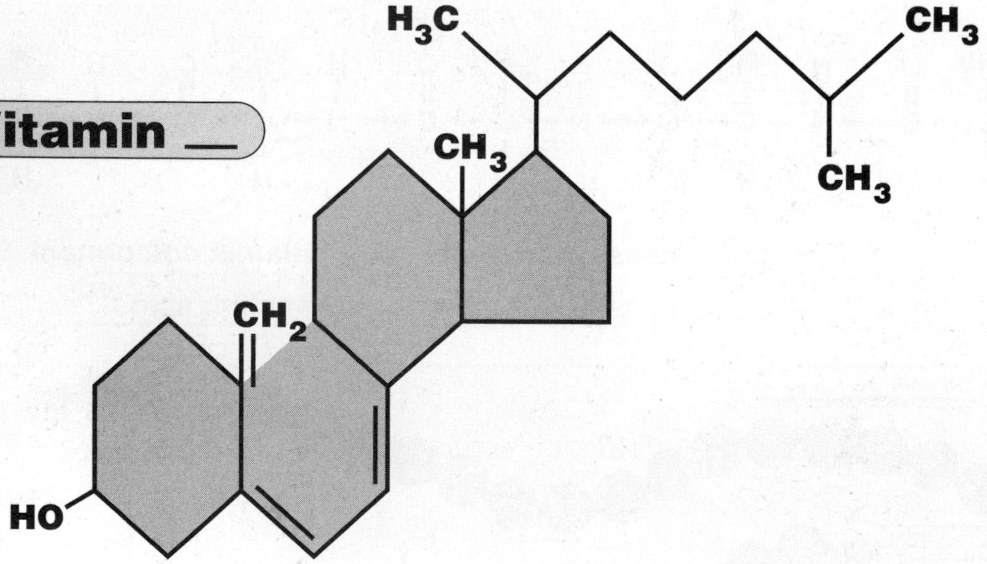

MINERALS

_____ Minerals

_____ minerals are required in amounts in excess of ____ mg/day.

Name and Symbol	Sources	Functions
_____ ____	green leafy vegetables, milk, and shellfish	bone development, nerve and muscle function, blood clotting
_____ ____	table salt, fish, and dairy products	acid-base balance, water balance, HCl formation
_____ ____	green leafy vegetables, fish, and cereals	bone development, nerve and muscle function, constituent of coenzymes
_____ ____	fish, poultry, meats, and dairy products	bone development, nerve and muscle function, buffer systems, enzyme component, energy transfer
_____ ____	Fruits, nuts, meats, and dairy products	nerve and muscle function
_____ ____	table salt, fish, and dairy products	nerve and muscle function, buffer system, electrolyte balance
_____ ____	fish, poultry, beans, eggs, cheese, and beef	hormone component, vitamin component, ATP production

_____ Minerals

_____ minerals are required in relatively small amounts: less than ____ mg/day.

_____ Cu	_____ I	_____ Se	_____ Zn
_____ Co	_____ Fe	_____ Si	
_____ Cr	_____ Mn	_____ Sn	
_____ F	_____ Mo	_____ V	

87

GASTROINTESTINAL TRACT

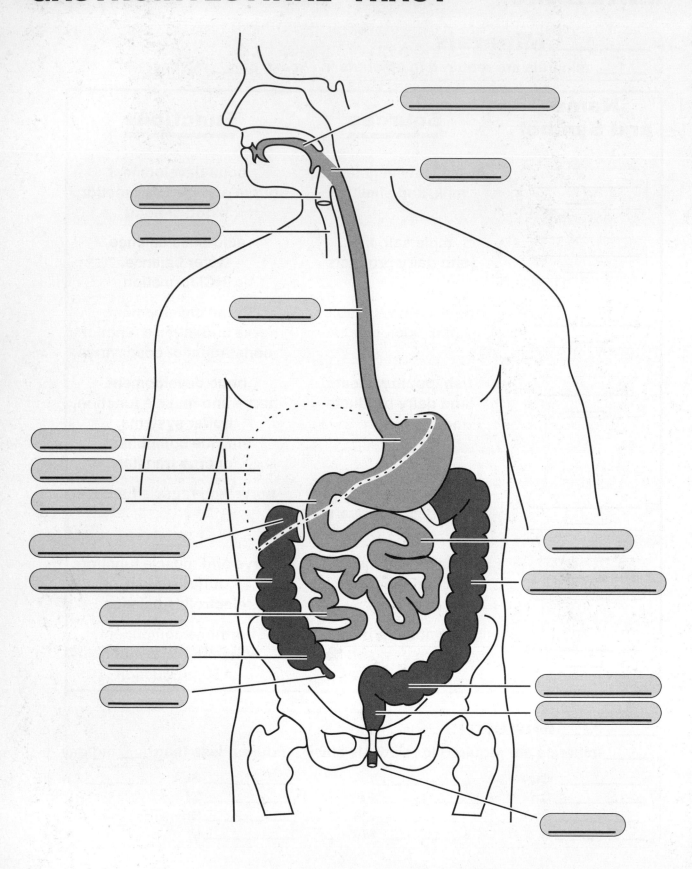

LAYERS OF THE GI TRACT

The wall of the GI tract from the esophagus to the anus has the same basic arrangement of tissues.

The four layers of the GI tract from the inside out are :
(1) _____   (3) _____
(2) _____   (4) _____

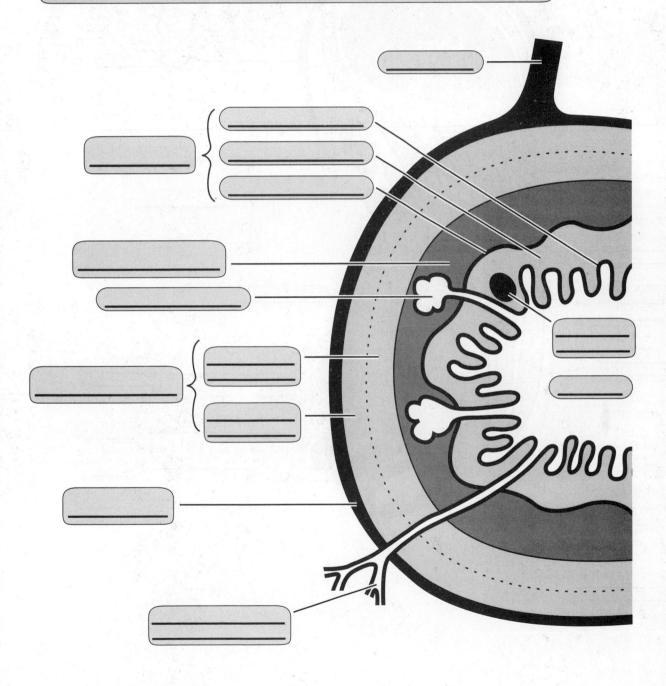

PERITONEUM

Midsagittal Section

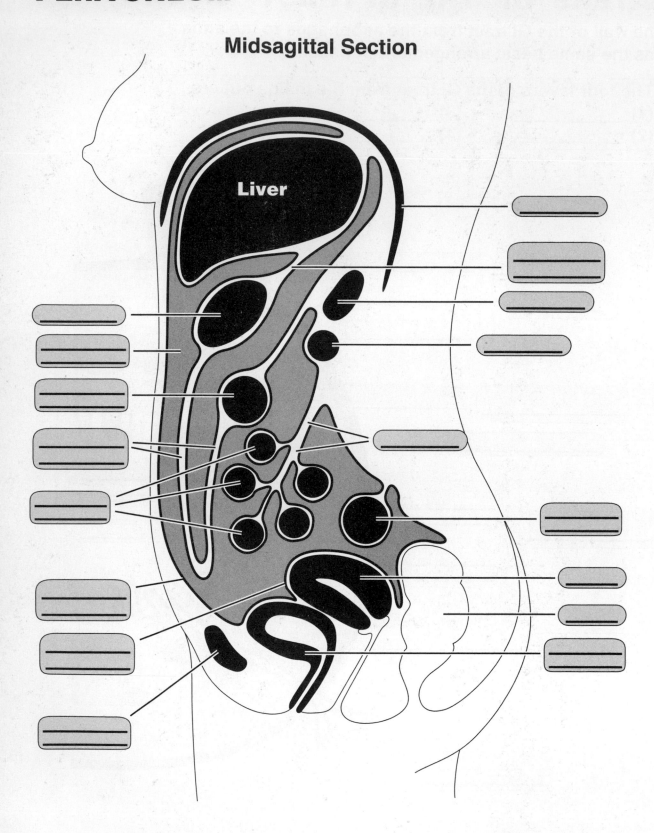

Liver

MOUTH (Oral Cavity)

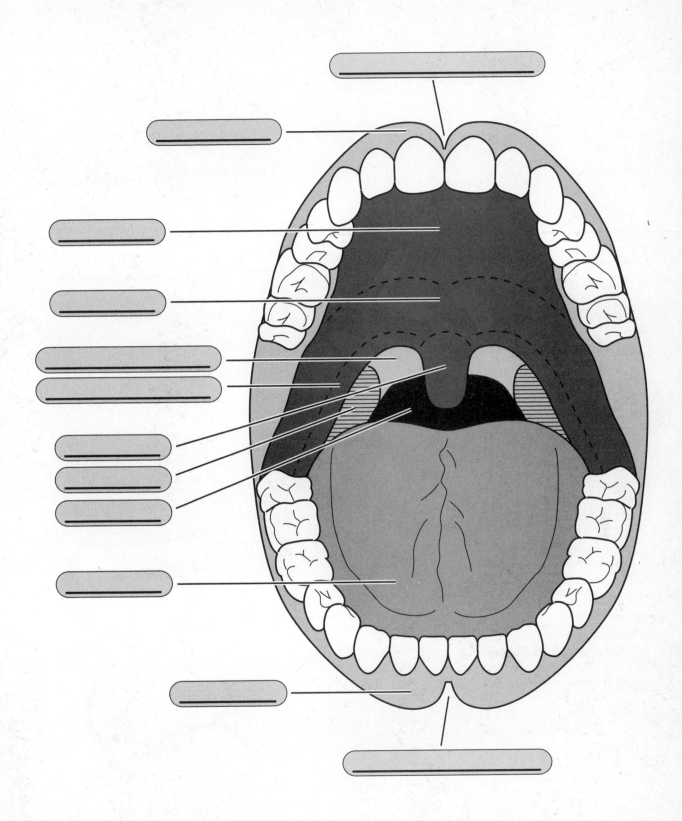

SALIVARY GLANDS

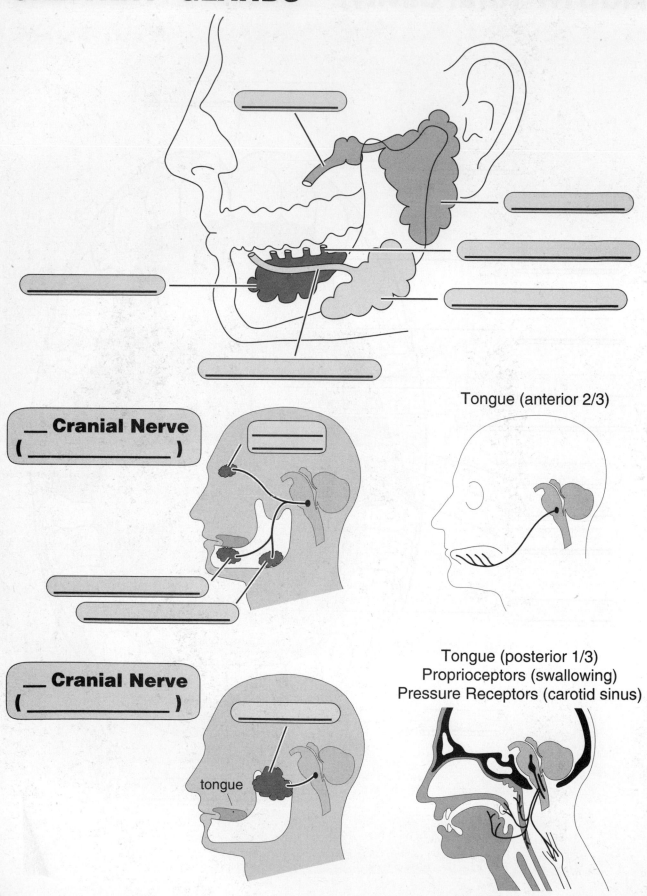

Tongue (anterior 2/3)

__ Cranial Nerve
(_____)

__ Cranial Nerve
(_____)

tongue

Tongue (posterior 1/3)
Proprioceptors (swallowing)
Pressure Receptors (carotid sinus)

TONGUE

Dorsum of the Tongue

root of tongue

Taste Zones :

(inside dotted line)

apex of tongue

Papillae : (locations)

Taste Bud

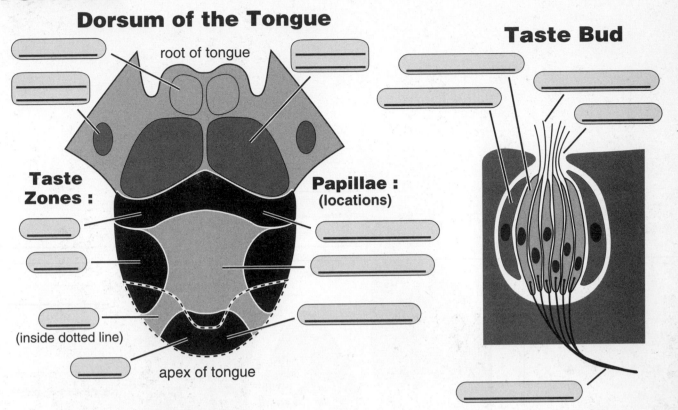

Circumvallate Papilla

upper surface of tongue

(lead to 9th cranial nerve)

TOOTH ANATOMY

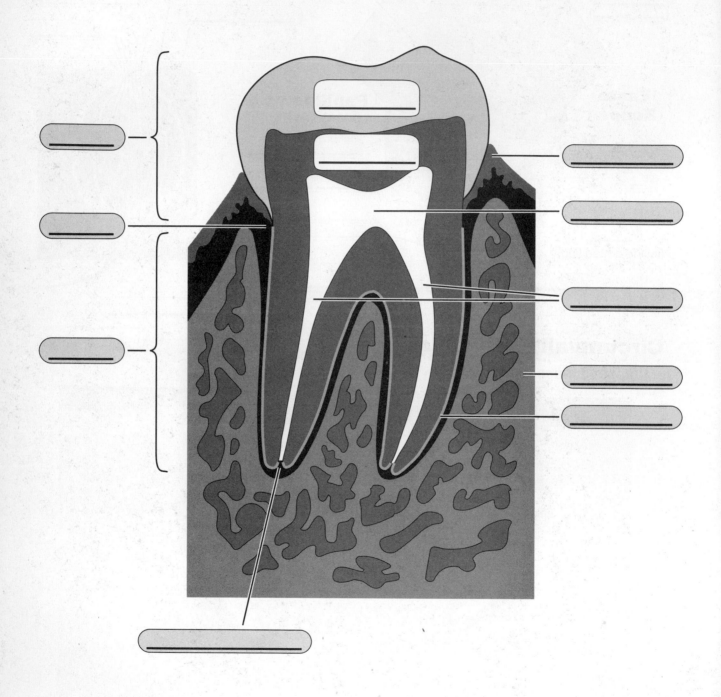

PERMANENT TEETH (32 Teeth)

_Incisors, _Cuspids, _Premolars, and _Molars

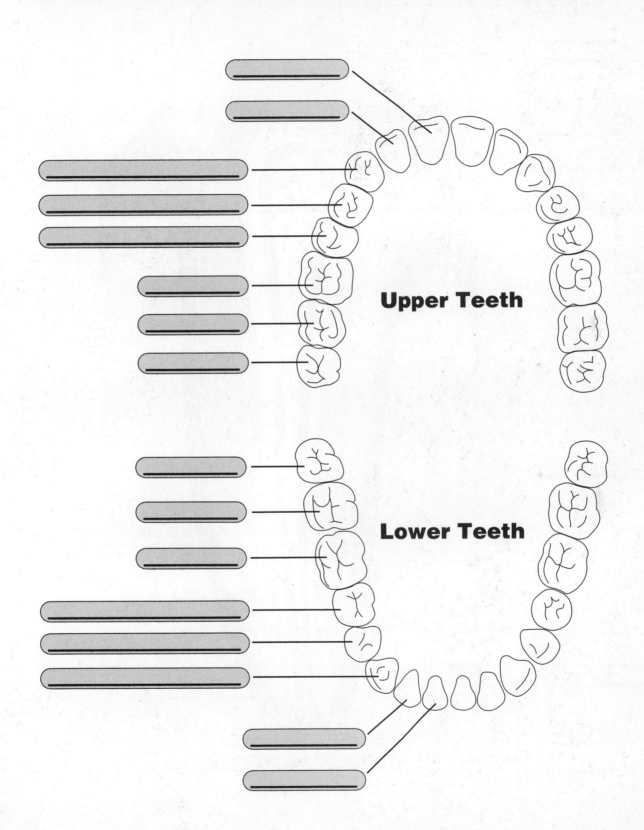

Upper Teeth

Lower Teeth

PREMOLAR WITH BLOOD VESSELS

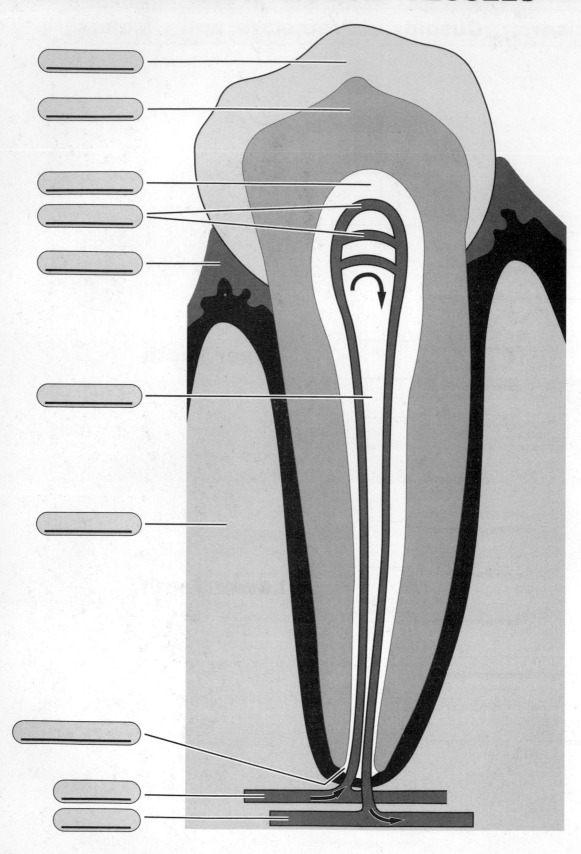

SWALLOWING (DEGLUTITION)

Stage 1

Stages 2 and 3

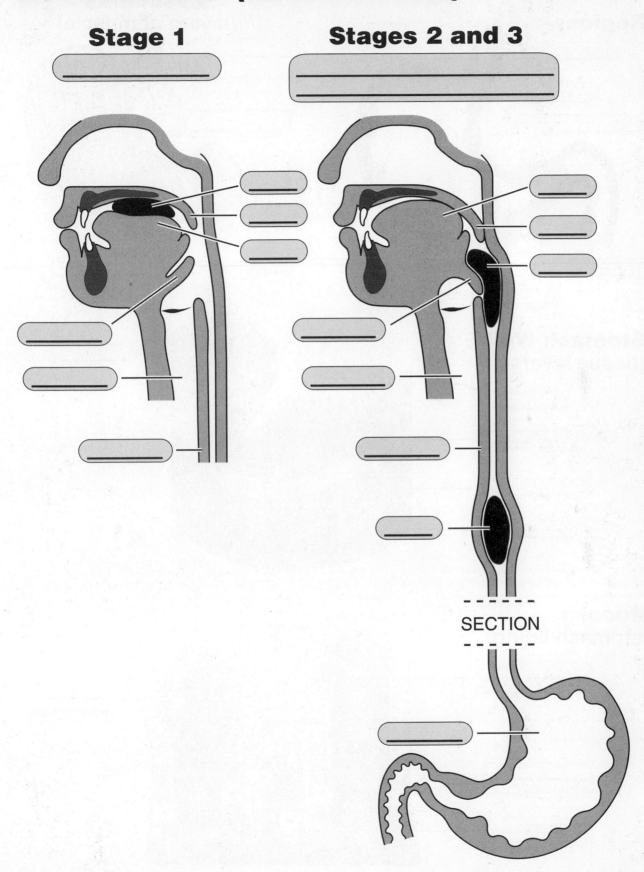

SECTION

STOMACH

Regions

Muscularis
(layers of muscle)

Stomach Wall
(tissue layers)

Mucosa
(stomach lining)

secretes _____

secretes _____

secretes _____

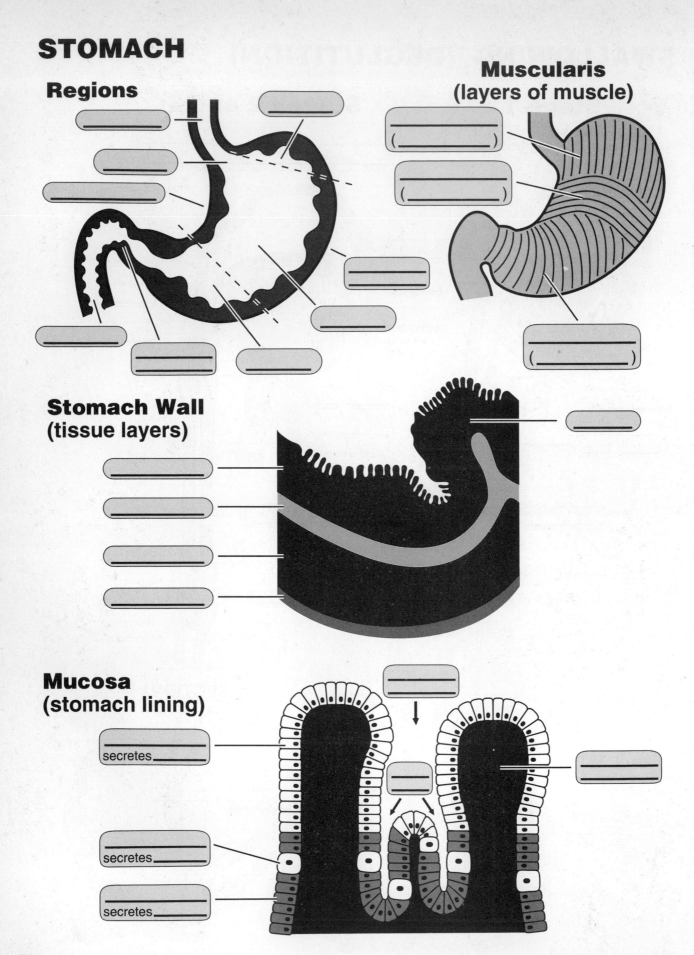

98

SMALL INTESTINE
Cross Section of the Small Intestine (diagramatic)

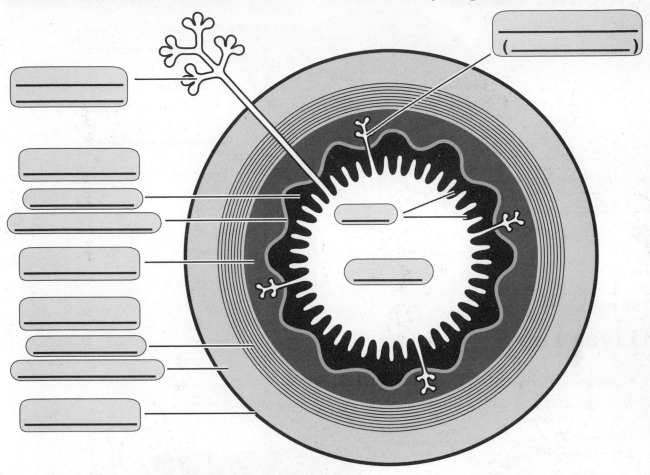

Lining (Mucosa and Submucosa)

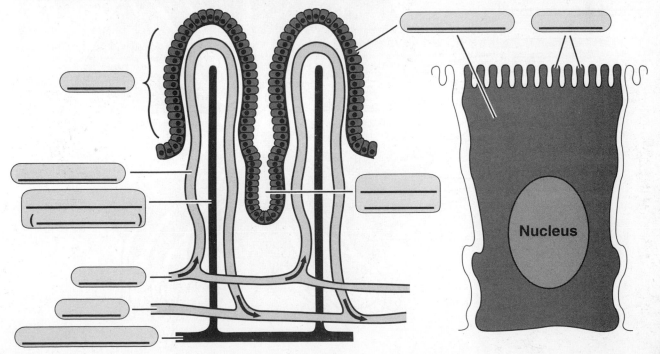

Nucleus

LIVER

Hepatic Ducts

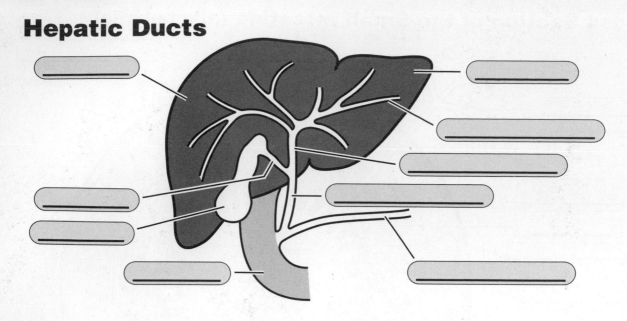

Liver Lobule

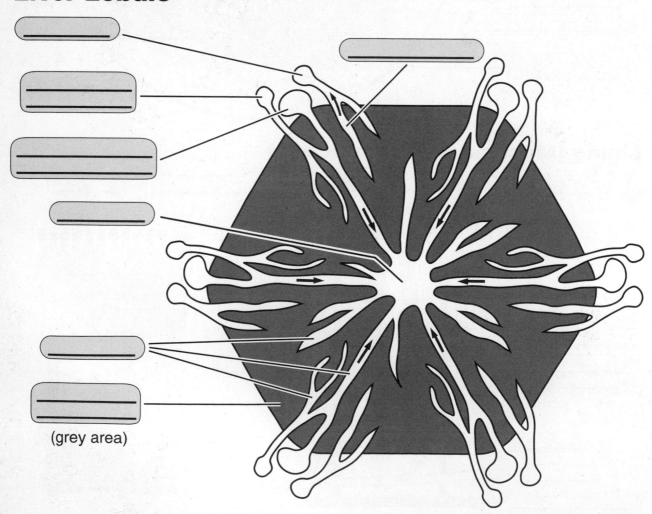

(grey area)

100

LIVER LOBULE

Blood Supply of a Liver Lobule

(For clarity, only the veins have been illustrated. Branches of the hepatic artery run parallel to branches of the hepatic portal vein.)

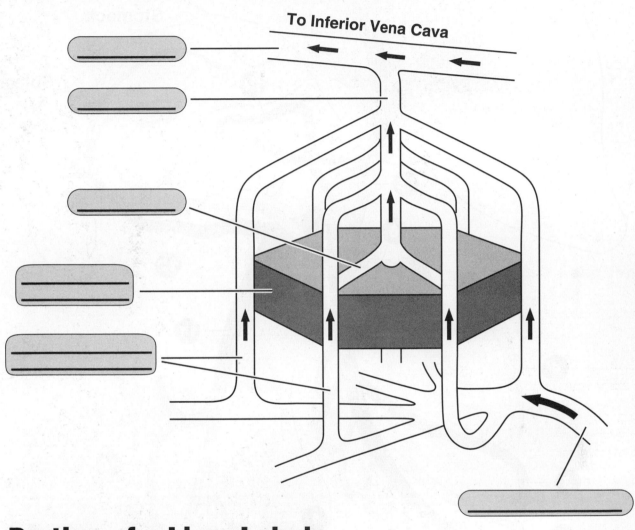

To Inferior Vena Cava

Portion of a Liver Lobule

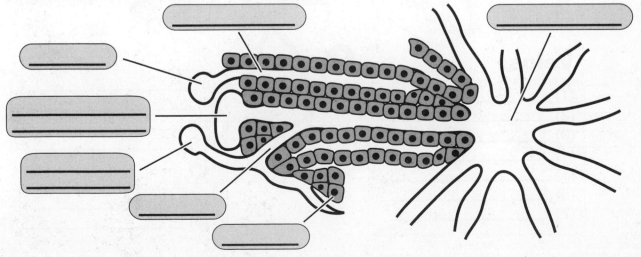

HEPATIC PORTAL SYSTEM

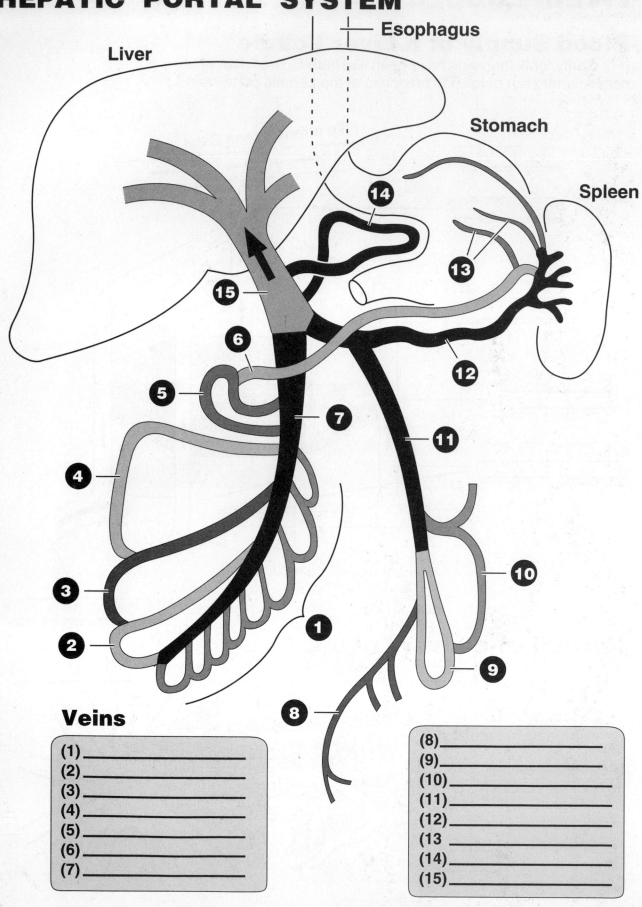

Esophagus

Liver

Stomach

Spleen

Veins

(1) _____
(2) _____
(3) _____
(4) _____
(5) _____
(6) _____
(7) _____

(8) _____
(9) _____
(10) _____
(11) _____
(12) _____
(13 _____
(14) _____
(15) _____

PANCREAS

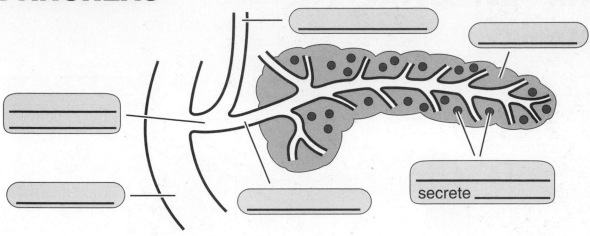

_____

_____

_____

_____

_____

_____ secrete _____

Exocrine Portion

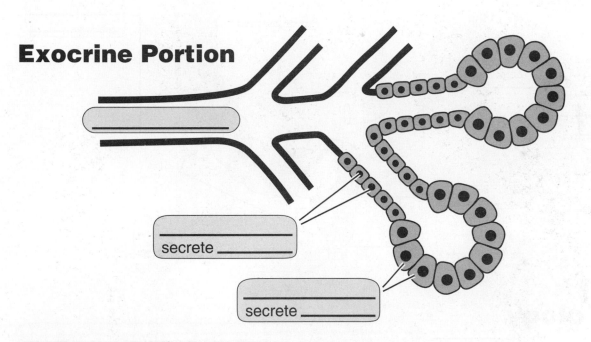

_____

_____ secrete _____

_____ secrete _____

Endocrine Portion

_____

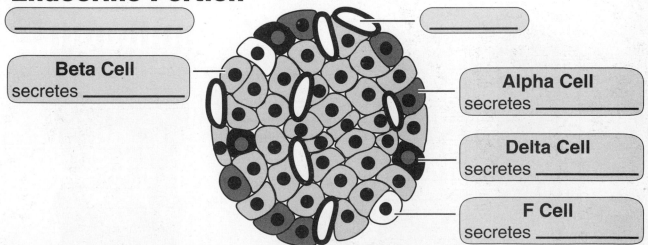

Beta Cell
secretes _____

Alpha Cell
secretes _____

Delta Cell
secretes _____

F Cell
secretes _____

LARGE INTESTINE
Anatomy

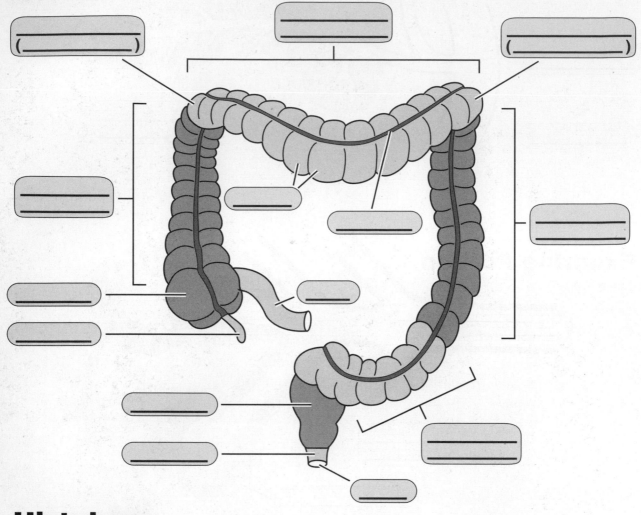

Histology

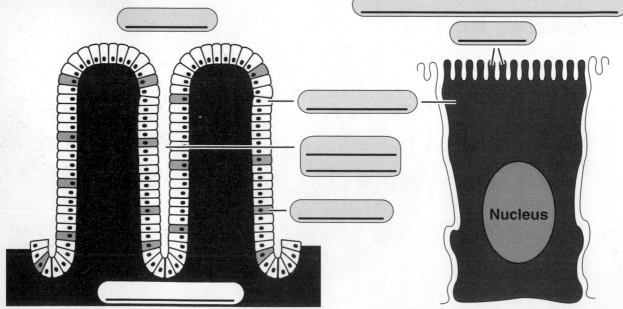

Nucleus

104

ENZYME ACTION

Example : hydrolysis of _____ (cane sugar).
The action of _____ (enzyme) on _____ (substrate).

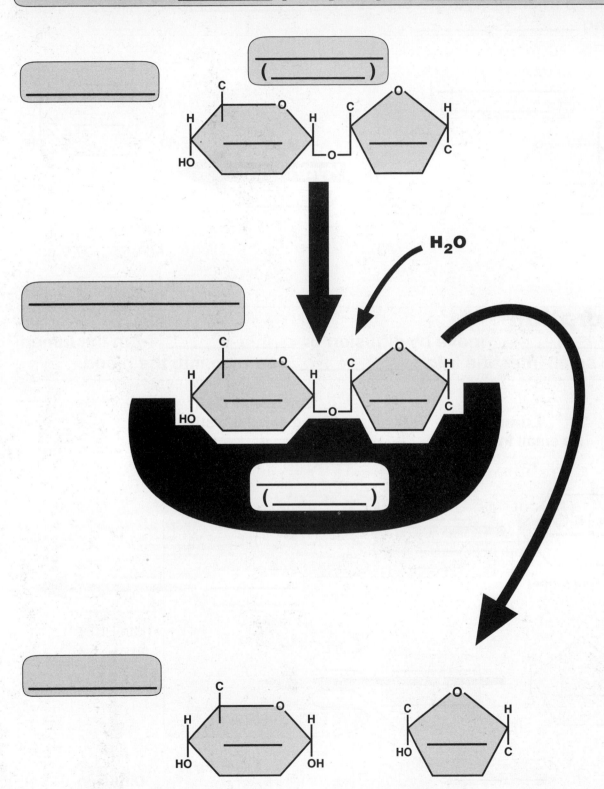

H₂O

105

CARBOHYDRATES

Digestion

Final step in carbohydrate digestion : _____ are split, forming _____ .

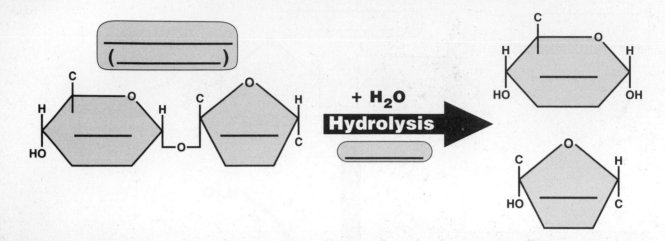

$+ H_2O$
Hydrolysis

Absorption

_____ move by diffusion or_____ from the lumen of the small intestine into _____ and then into the blood.

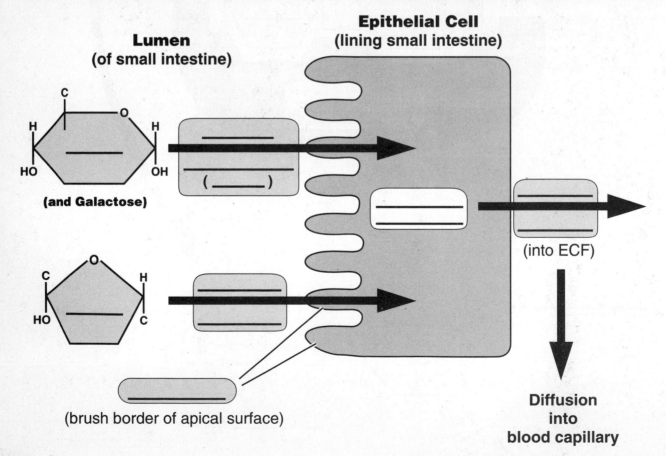

Lumen
(of small intestine)

Epithelial Cell
(lining small intestine)

(and Galactose)

(into ECF)

(brush border of apical surface)

Diffusion into blood capillary

LIPIDS

Digestion

Large _____ must be broken down into smaller droplets and coated with _____ (emulsification) before they can be efficiently digested by the enzyme lipase. _____ splits two fatty acids from a _____ , forming 1 _____ and 2 fatty acids.

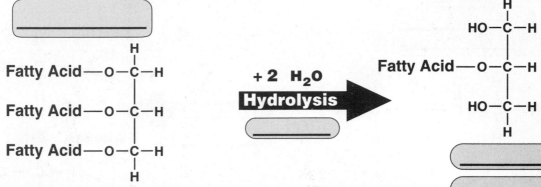

Fatty Acid—O—C—H

Fatty Acid—O—C—H

Fatty Acid—O—C—H

+ 2 H₂O
Hydrolysis

Fatty Acid—O—C—H

HO—C—H

HO—C—H

Absorption

_____ and _____ combine with bile salts, forming tiny water-soluble spheres called _____ . As they are released from the micelles, fatty acids and monoglycerides move by _____ from the lumen of the small intesine into _____ cells.

Inside the epithelial cells _____ are resynthesized, forming spherical masses called _____ , which leave the epithelial cells by exocytosis and enter _____ capillaries (lacteals).

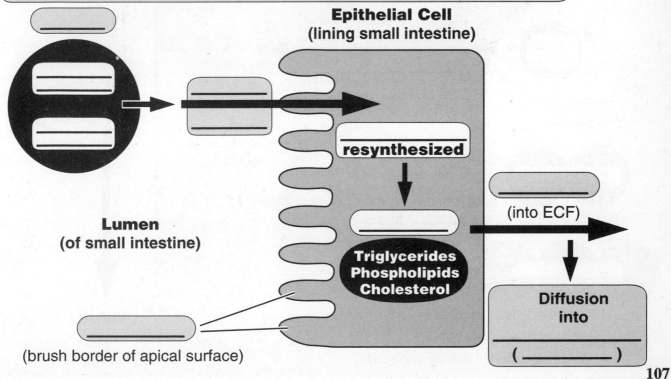

Epithelial Cell
(lining small intestine)

resynthesized

(into ECF)

**Triglycerides
Phospholipids
Cholesterol**

**Diffusion
into**

(_____)

Lumen
(of small intestine)

(brush border of apical surface)

PROTEINS

Digestion

Final step in protein digestion : _____ are split, forming _____ ; or amino acids are cleaved from the _____ or _____ end of a peptide.

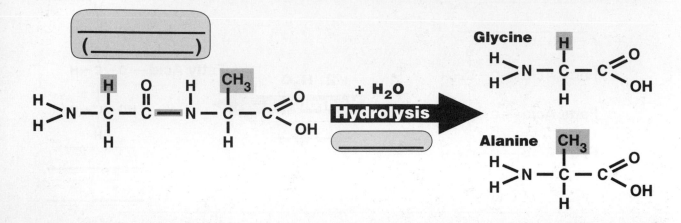

Absorption

_____ , _____ , and _____ move by _____ active transport from the lumen of the small intestine into _____ cells and then into the _____ .

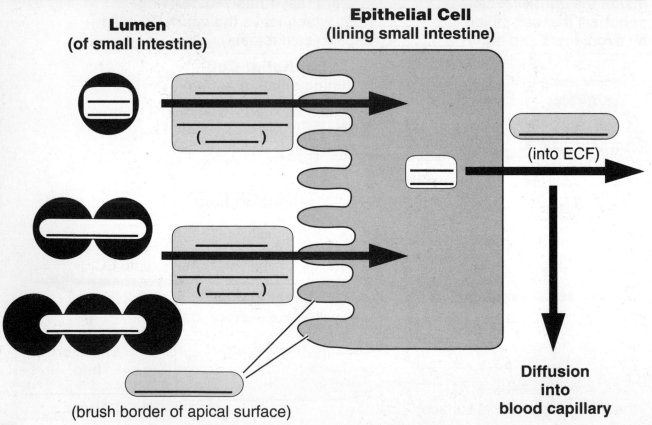

108

VITAMINS, MINERALS, and WATER

Digestion
Vitamins are small _____ molecules; minerals are _____ molecules
that break apart into _____ when dissolved in water. Vitamins and minerals
do not have to be digested, since they are already _____ enough to pass through
the _____ cells that line the small intestine.

Absorption
Water-soluble vitamins are absorbed by _____ .
Fat-soluble vitamins are absorbed by the same mechanism as _____.
Minerals are absorbed by _____ or _____ .
Most water absorption occurs in the small intestine by _____ .

Fluid Balance in the Gastrointestinal Tract

Fluids Ingested and Secreted   **Fluids Absorbed**

Ingested Liquids = __liters

Saliva = __liter

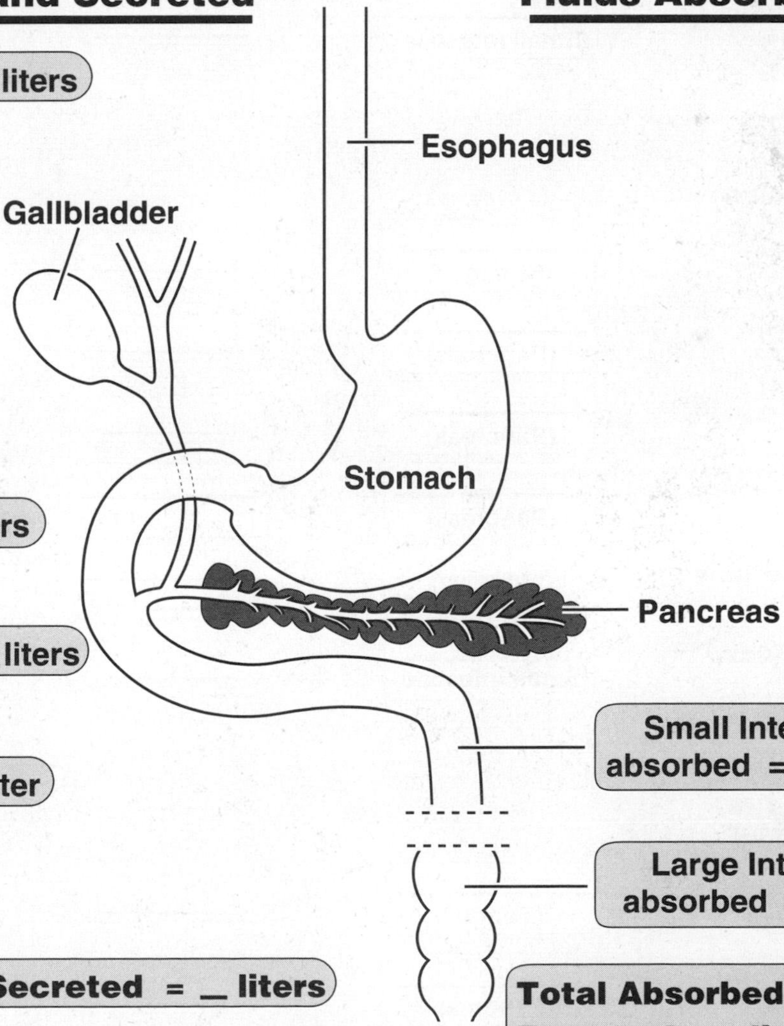

                                              — Esophagus

Gallbladder

                                              Stomach

Bile = __liter

Gastric Juice = __liters

                                              — Pancreas

Pancreatic Juice = __liters

                                              Small Intestine
                                              absorbed = __liters

Intestinal Juice = __liter

                                              Large Intestine
                                              absorbed = __liter

Total Ingested or Secreted = __ liters

                                              Total Absorbed = __liters
                                              Excreted = __liter

DIGESTIVE ENZYMES

Enzyme	Source	Substrate	Products
CARBOHYDRATE – DIGESTING ENZYMES			
Salivary Amylase	_____	_____	_____
Pancreatic Amylase	_____ (Pancreas)	_____	_____
Alpha-Dextrinase	_____ (Small Intestine)	_____	_____
Maltase	_____ (Small Intestine)	_____	_____
Sucrase	_____ (Small Intestine)	_____	_____
Lactase	_____ (Small Intestine)	_____	_____
PROTEIN – DIGESTING ENZYMES			
Pepsin	_____ (Stomach)	_____	_____
Trypsin	_____ (Pancreas)	_____	_____
Chymotrypsin	_____ (Pancreas)	_____	_____
Carboxypeptidase	_____ (Pancreas)	(_____ End)	_____
Aminopeptidase	_____ (Small Intestine)	(_____ End)	_____
Dipeptidase	_____ (Small Intestine)	_____	_____
LIPID – DIGESTING ENZYMES			
Lingual Lipase	_____ (Tongue)	_____	_____
Pancreatic Lipase	_____ (Pancreas)	_____	_____

HORMONAL CONTROL OF DIGESTION

Gastrin : secretion stimulated by _____ and _____ in the stomach; stimulates the secretion of _____ (HCl and pepsinogen).

Secretin : secretion stimulated by _____ in the duodenum; enhances the flow of _____ rich in _____ (HCO_3^-) from the liver.

CCK : secretion stimulated by _____ and _____ in the duodenum; stimulates the secretion of _____ and pancreatic _____ .

GIP : secretion stimulated by _____ and _____ in the duodenum; inhibits the secretion of _____ .

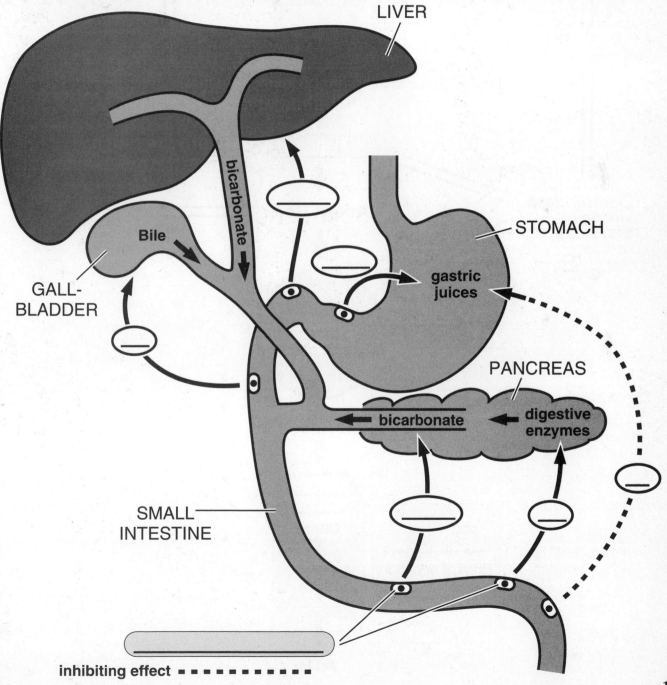

LIVER

bicarbonate

Bile

STOMACH

GALL-
BLADDER

gastric
juices

PANCREAS

bicarbonate

digestive
enzymes

SMALL
INTESTINE

inhibiting effect ▪ ▪ ▪ ▪ ▪ ▪ ▪ ▪ ▪ ▪

111

METABOLISM OVERVIEW

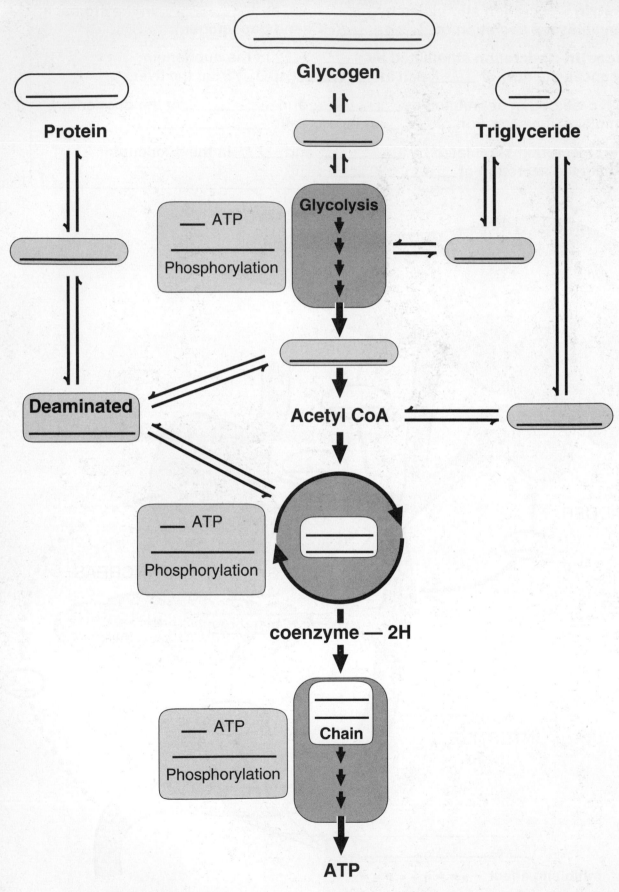

CARBOHYDRATE METABOLISM

_____ :
the synthesis of glucose from noncarbohydrate precursors

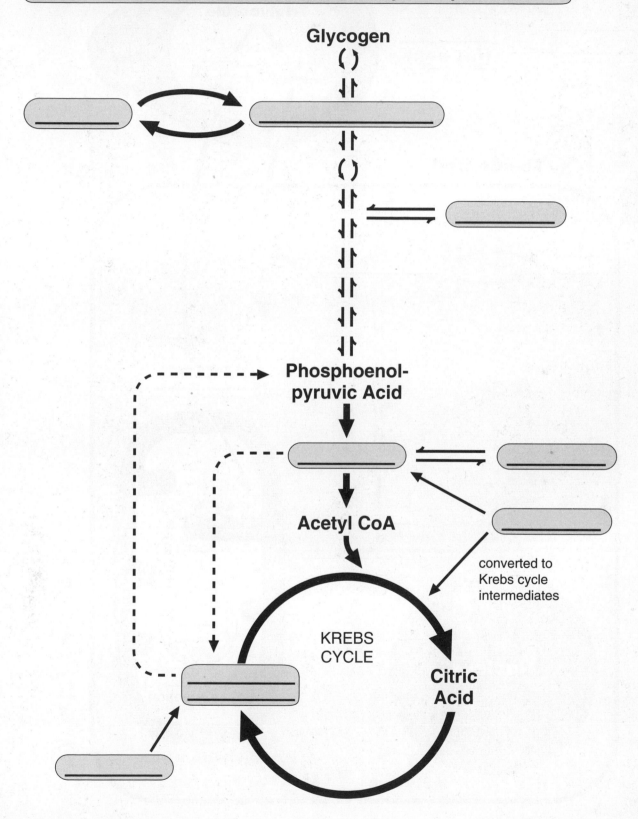

Glycogen

Phosphoenol-
pyruvic Acid

Acetyl CoA

KREBS
CYCLE

Citric
Acid

converted to
Krebs cycle
intermediates

113

LIPID METABOLISM

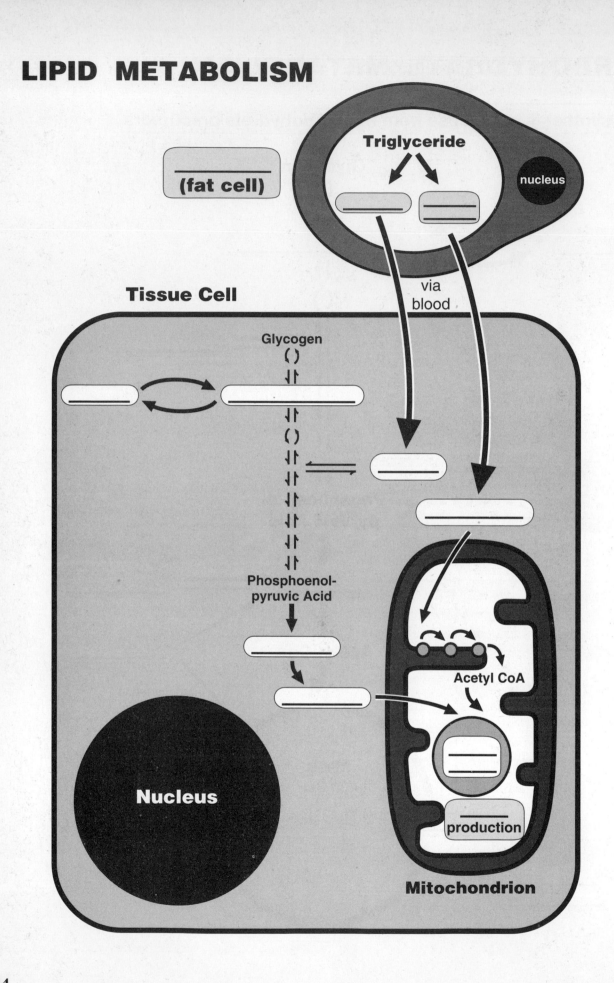

_____
(fat cell)

Triglyceride

nucleus

Tissue Cell

via
blood

Glycogen
( )

( )

Phosphoenol-
pyruvic Acid

Acetyl CoA

Nucleus

production

Mitochondrion

ATHEROSCLEROSIS

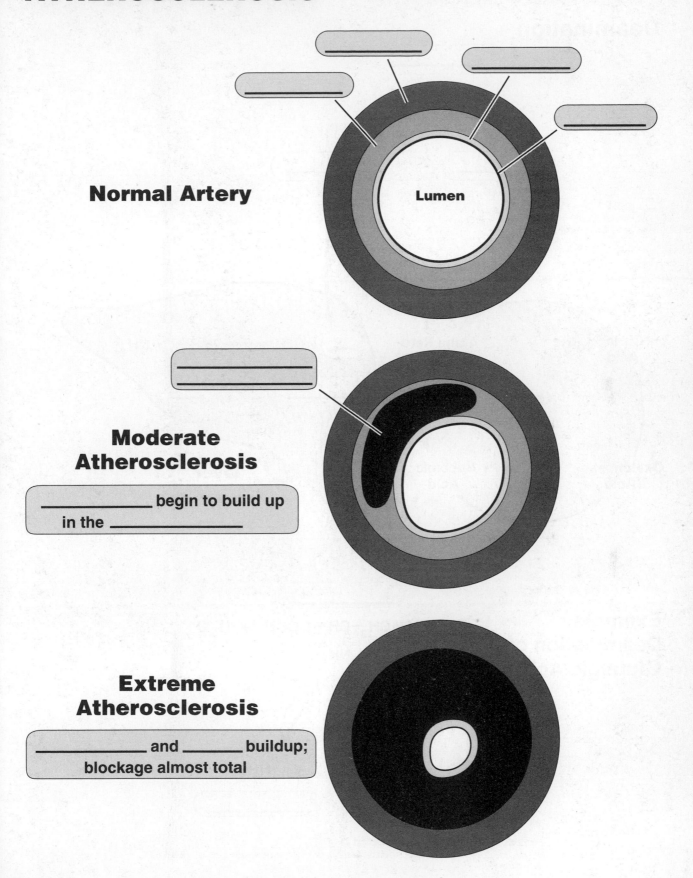

Normal Artery

Lumen

**Moderate
Atherosclerosis**

_____ begin to build up
in the _____

**Extreme
Atherosclerosis**

_____ and _____ buildup;
blockage almost total

PROTEIN METABOLISM

Deamination

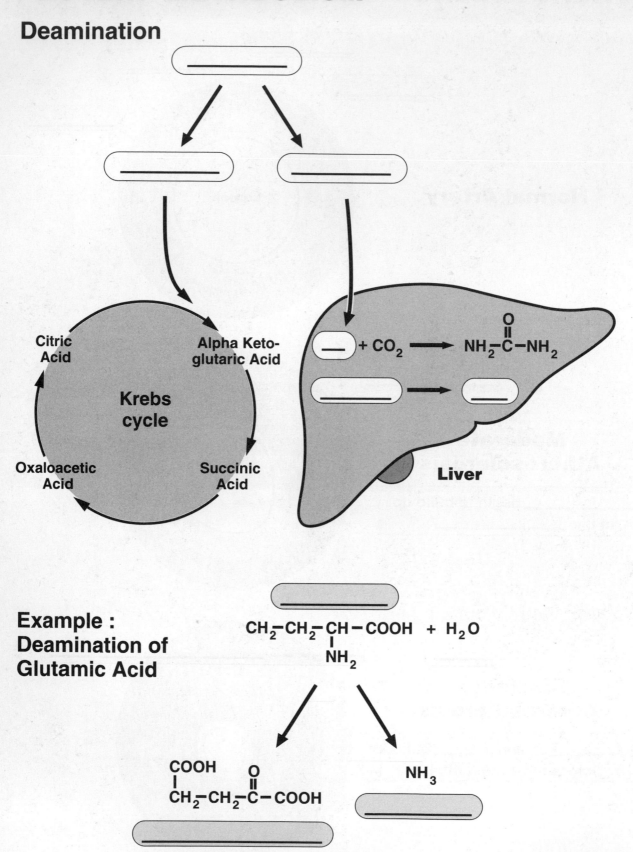

Citric Acid

Alpha Keto-glutaric Acid

Krebs cycle

Oxaloacetic Acid

Succinic Acid

$$NH_2 - \overset{\overset{\displaystyle O}{\|}}{C} - NH_2$$

+ CO_2 →

Liver

**Example :
Deamination of
Glutamic Acid**

$$CH_2 - CH_2 - \underset{\underset{\displaystyle NH_2}{|}}{CH} - COOH \ + \ H_2O$$

$$\underset{\underset{\displaystyle CH_2-CH_2-\overset{\overset{\displaystyle O}{\|}}{C}-COOH}{\displaystyle COOH}}{}$$

NH_3

ALCOHOL BIOTRANSFORMATION

Breakdown of Alcohol by Liver Cells

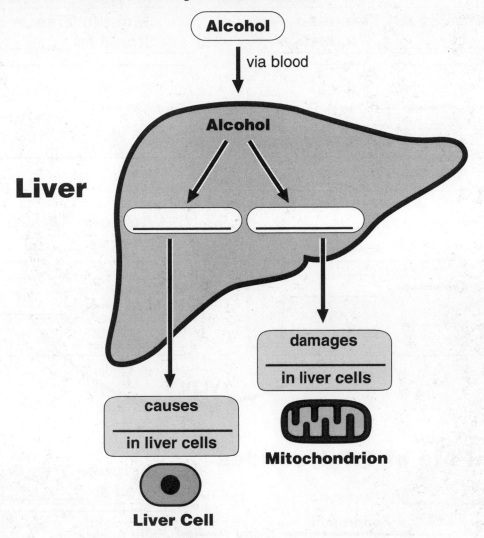

Chronic Overuse of Alcohol

Stage 1 : _____

Stage 2 : _____

Before Liver Damage

chronic overuse
stimulates

_____

↓

_____
rate of catabolism
of alcohol

After Liver Damage

decreased number
of functioning

_____

↓

_____
rate of catabolism
of alcohol

ABSORPTIVE (FED) STATE

The Fate of Glucose

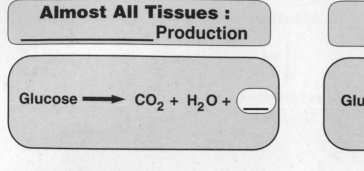

Almost All Tissues :
_____ Production

Glucose $\longrightarrow$ $CO_2 + H_2O +$ ____

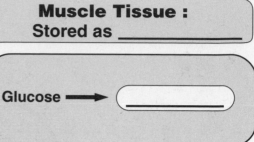

Muscle Tissue :
Stored as _____

Glucose $\longrightarrow$ _____

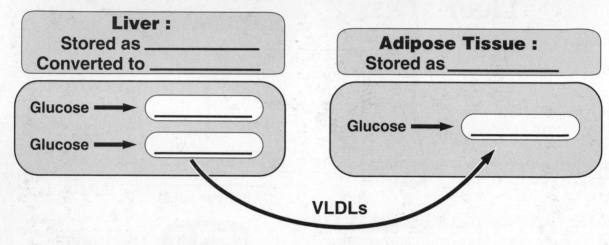

Liver :
Stored as _____
Converted to _____

Glucose $\longrightarrow$ _____
Glucose $\longrightarrow$ _____

Adipose Tissue :
Stored as _____

Glucose $\longrightarrow$ _____

VLDLs

The Fate of Triglycerides

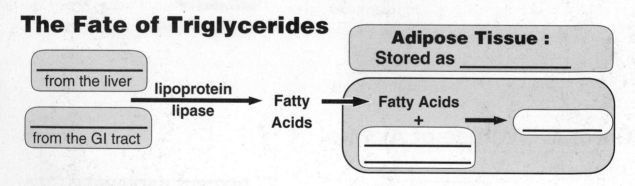

_____
from the liver

_____
from the GI tract

lipoprotein
lipase $\longrightarrow$

Fatty
Acids $\longrightarrow$

Adipose Tissue :
Stored as _____

Fatty Acids
+
_____ $\longrightarrow$ _____

The Fate of Amino Acids

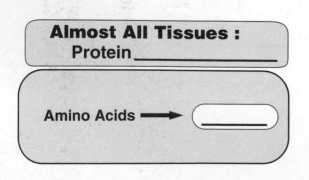

Almost All Tissues :
Protein_____

Amino Acids $\longrightarrow$ _____

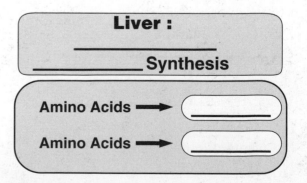

Liver :
_____
_____ Synthesis

Amino Acids $\longrightarrow$ _____
Amino Acids $\longrightarrow$ _____

118

POSTABSORPTIVE (FASTING) STATE
Maintenance of the Normal Blood Glucose Level
(___ to ___ mg/ml of Blood)

Glucose Synthesis

Neurons require glucose for _____ (they also can use
_____ during starvation). During the postabsorptive state
glucose is synthesized by the liver from _____ and several
noncarbohydrate precursors.

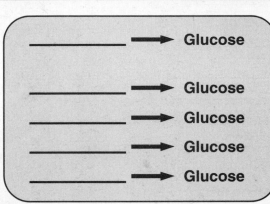

Liver :

_____
(glycogen breakdown)

_____
(glucose synthesis from
noncarbohydrates)

_____ ⟶ Glucose

_____ ⟶ Glucose

_____ ⟶ Glucose

_____ ⟶ Glucose

_____ ⟶ Glucose

Glucose-Sparing Reactions

During the postabsorptive state all body cells (except neurons)
reduce their_____ and switch over to _____ as their
main source of ATP. This "spares" the available _____ for utilization
by neurons.

Adipose Tissue :

_____
(_____ breakdown)

_____ ⟶ Fatty Acids
+
Glycerol

**Almost All Tissues
(except nervous) :**
_____ Production

_____ ⟶ CO_2 + H_2O + ATP

_____ ⟶ CO_2 + H_2O + ATP

Liver :
Conversions

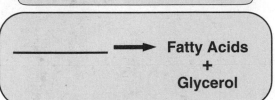

Fatty Acids ⟶ _____

119

HORMONES THAT AFFECT METABOLISM

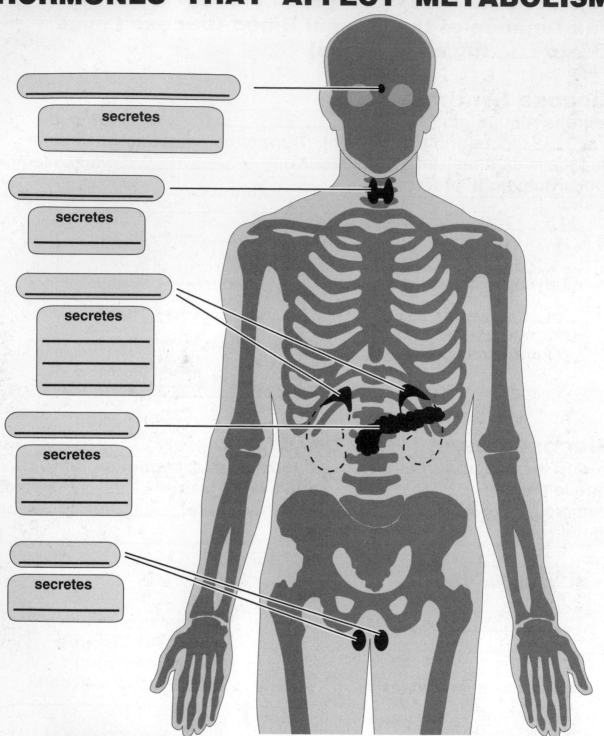

_____
secretes
_____

_____
secretes
_____

_____
secretes
_____
_____

_____
secretes
_____
_____

_____
secretes
_____

Metabolic Functions:

Human Growth Hormone : _____ ; _____ ; protein synthesis.
Thyroid Hormones : _____ ; _____ synthesis; increase _____ .
Epinephrine : glycogenolysis; _____ ; increases _____ .
Norepinephrine : _____ .
Cortisol : gluconeogenesis; _____ ; _____ synthesis.
Insulin : glycogenesis; _____ ; protein synthesis.
Glucagon : _____ .
Testosterone : increases _____ .

Part III : Terminology

Pronunciation Guide

Pronunciation Key

Accented Syllables

The strongest accented syllable is in capital letters : **dī - ag - NŌ - sis**

Secondary accents are indicated by a prime (′): **fiz′ - ē - OL - ō - jē**

Vowels

Long vowels are indicated by a horizontal line (macron) above the letter :

ā as in **make** **ē** as in **be** **ī** as in **ivy** **ō** as in **pole**

Short vowels are unmarked :

e as in **pet** **i** as in **pit** **o** as in **pot** **u** as in **bud**

Other Phonetic Symbols

a as in **about** **oo** as in **do** **yoo** as in **cute** **oy** as in **noise**

acetaldehyde **as′ - et - AL - de - hīd′**
acetyl CoA **AS - e - til kō - Ā**
acetyl coenzyme A **AS - e - til kō - EN - zīm Ā**
acinar **AS - i - nar**
acini **AS - i - nē**
adenosine **a - DEN - ō - sēn**
adipose **AD - i - pōs**
adrenaline **a - DREN - a - lin**
adventitia **ad - ven - TISH - ya**
alimentary **al′ - i - MEN - tar - ē**
alpha **AL-fa**
alveolar **al - VĒ - ō - lar**
alveolus **al - VĒ - ō - lus**
amino **a - MĒ - nō**
aminopeptidase **a - mē′ - nō - PEP - tid - ās**
ampulla **am - POOL - la**
amylase **AM - i - lās**
anabolism **a - NAB - ō - lizm**
anal **Ā - nal**
anion **AN - ī - on**
antrum **AN - trum**
anus **Ā - nus**
apical **AP - i - kal**
apoferritin **ap′ - ō - FER - i - tin**
apoprotein **ap′ - ō - PRŌ - tēn**
appendage **a - PEN - dij**
areolar **a - RĒ - ō - lar**
ascorbic **a - SKOR - bik**
Auerbach **OW - er - bak**
autonomic **aw′ - tō - NOM - ik**

barbiturate **bar - BICH - er - it**
basal **BĀ - sal**
beta **BĀ - ta**
bicuspid **bī - KUS - pid**
bile **BĪL**
biliary **BIL - ē - er - ē**
bilirubin **bil′ - ē - ROO - bin**
biliverdin **bil′ - ē - VER - din**
biotin **BĪ - ō - tin**
bolus **BŌ - lus**
Brunner's **BRUN - erz**
buccal **BUK - al**

canaliculi **kan′ - a - LIK - yoo - lī**
canaliculus **kan′ - a - LIK - yoo - lus**
canine **KĀ - nīn**
carbohydrate **kar′ - bō - HĪ - drāt**
carboxyl **kar - BOK - sil**
carboxypeptidase **kar - bok′ - sē - PEP - ti - dās**
cardia **KAR - dē - a**
carotene **KAR - ō - tēn**
catabolism **ka - TAB - ō - lizm**
cation **KAT - ī - on**
caudate **KAW - dāt**
cecum **SĒ - kum**
cellulose **SEL - yoo - lōs**
cementum **se - MEN - tum**
cephalic **se - FAL - ik**
cholecystokinin **kō′ - lē - sis′ - tō - KĪN - in**
cholesterol **kō - LES - ter - ol**
chylomicron **kī - lō - MĪK - ron**
chyme **KĪM**
chymotrypsin **kī′ - mō - TRIP - sin**
chymotrypsinogen **kī′ - mō - trip - SIN - ō - jen**
circumvallate **ser′ - kum - VAL - āt**

cobalamin kō - BAL - a - min
coenzyme kō - EN - zīm
colic KOL - ik
colon KŌ - lon
cortisol KOR - ti - sol
crypt KRIPT
cuspid KUS - pid
cystic SIS - tik

deamination dē - am′ - i - NĀ - shun
deciduous di - SID - yoo - us
defecation def - e - KĀ - shun
deglutition dē - gloo - TI - shun
dentes DEN - tēz
dentin DEN - tin
dentition den - TI - shun
deoxyribonuclease dē - ok′ - sē - rī′ - bō - NOO - klē - ās
deoxyribonucleic dē - ok′ - sē - rī′ - bō - noo - KLĀ - ik
dextrinase DEKS - trin - ās
dipeptidase dī - PEP - ti - dās
dipeptide dī - PEP - tīd
disaccharide dī - SAK - a - rīd
diverticulum dī - ver - TIK - yoo - lum
duodenal doo′ - ō - DĒ - nal
duodenum doo′ - ō - DĒ - num

electrolyte e - LEK - trō - līt
emulsification ē - mul′ - si - fi - KĀ - shun
endocrine EN - dō - krin
endodontics en′ - dō - DON - tiks
endopeptidase en′ - dō - PEP - ti - dās
enteric en - TER - ik
enteroendocrine en′ - ter - ō - EN - dō - krin
enterogastric en′ - ter - ō - GAS - trik
enterokinase en′ - ter - ō - KĪ - nās
enzyme En - zīm
epiglottis ep′ - i - GLOT - is
epinephrine ep′ - i - NEF - rin
epiploic ep′ - i - PLŌ - ik
epithelium ep′ - i - THĒ - lē - um
equilibrium ē′ - kwi - LIB - rē - um
esophageal e - sof′ - a - JĒ - al
esophagus e - SOF - a - gus
excrement EK - skra - ment
excretion ek - SKRĒ - shun
exocrine EK - sō - krin
exopeptidase ek′ - sō - PEP - ti - dās
extrinsic ek - STRIN - sik

facial FĀ - shul
falciform FAL - si - form
fauces FAW - sēs
fecal FĒ - kal
feces FĒ - sēz

ferritin FER - i - tin
filiform FIL - i - form
flatus FLĀ - tus
flexure FLEK - sher
folacin FŌL - a - sin
folic FŌ - lik
foramen fo - RĀ - men
frenulum FREN - yoo - lum
fructose FRUK - tōs
fundus FUN - dus
fungiform FUN - ji - form

galactose ga - LAK - tōs
gastric GAS - trik
gastrin GAS - trin
gastroenterology gas′ - trō - en′ - ter - OL - ō - jē
gastrointestinal gas′ - trō - in - TES - ti - nal
gingivae jin - JI - vē
glossopharyngeal glos′ - ō - fa - RIN - jē - al
glottis GLOT - is
glucagon GLOO - ka - gon
gluconeogenesis gloo′ - kō - nē′ - ō - JEN - e - sis
glucose GLOO - kōs
glyceraldehyde glis′ - er - AL - de - hīd
glycerol GLIS - er - ol
glycogen GLĪ - kō - jen
glycogenesis glī′ - kō - JEN - e - sis
glycogenolysis glī′ - kō - je - NOL - i - sis
glycolysis glī - KOL - i - sis

haustra HAWS - tra
haustral HAWS - tral
haustrum HAWS - trum
hepatic he - PAT - ik
hepatocyte HEP - a - tō - sīt
hepatopancreatic hep′ - a - tō - pan′ - krē - A - tik
hiatus hī - Ā - tus
hydrochloric hī′ - drō - KLOR - ik
hydrolysis hī - DROL - i - sis
hydrolytic hī - drō - LIT - ik
hypopharynx hī′ - pō - FAR - inks

ileocecal il′ - ē - ō - SĒ - kal
ileum IL - ē - um
incisor in - SĪ - zer
insulin IN - su - lin
intestinal in - TES - tin - al
intrinsic in - TRIN - sik
islet Ī - let

jejunum jē - JOO - num

keto KĒ - tō
ketogenesis kē′ - tō - JEN - e - sis

ketone KĒ-tōn
kilocalorie KIL-ō-kal'-ō-rē
Krebs KREBZ
Kuppfer's KOOP-ferz

labia LĀ-bē-a
labial LĀ-bē-al
lactase LAK-tās
lacteal LAK-tē-al
lactose LAK-tōs
lamina propria LAM-i-na PRŌ-prē-a
Langerhans LANG-er-hanz
laryngopharynx la-rin'-gō-FAR-inks
larynx LAR-inks
Lieberkühn LĒ-ber-kyoon
ligamentum teres lig'-a-MEN-tum TĒ-rēz
lingual LIN-gwal
lipase LĪ-pās
lipid LIP-id
lipogenesis lip'-ō-GEN-e-sis
lipolysis li-POL-i-sis
lipoprotein lip'-ō-PRŌ-tēn
lobule LOB-yool
lumen LOO-men
lymphatic lim-FAT-ik
lysozyme LĪ-sō-zīm

maltase MAWL-tās
maltose MAWL-tōs
mastication mas'-ti-KĀ-shun
Meissner's MĪS-nerz
mesentery MEZ-en-ter'-ē
mesoappendix mez'-ō-a-PEN-diks
mesocolon mez'-ō-KŌ-lon
mesothelium mez'-ō-THĒ-lē-um
metabolism me-TAB-ō-lizm
metabolite me-TAB-ō-līt
micelle mī-SEL
microvilli mī'-krō-VIL-ī
microvillus mī'-krō-VIL-us
monoglyceride mon'-ō-GLIS-er-īd
monosaccharide mon'-ō-SAK-a-rīd
monounsaturated mon'-ō-un-SA-chur-ā-ted
motility mō-TIL-i-tē
mucin MYOO-sin
mucosa myoo-KŌ-sa
mucosae myoo-KŌ-sē
mucous myoo-kus
mucus myoo-kus
muscularis MUS-kyoo-la'-ris
myenteric mī'-en-TER-ik

nasopharynx nā'-zō-FAR-inks
niacin NĪ-a-sin

nicotinic nik'-ō-TIN-ik
noradrenaline nor'-a-DREN-a-lin
norepinephrine nor'-ep-i-NEF-rin
nuclease NOO-klē-ās
nucleic noo-KLĀ-ik
nucleosidase noo'-klē-ō-SĪ-dās
nucleotide NOO-klē-ō-tīd
nutrient NOO-trē-ent

Oddi OD-ē
omentum ō-MENT-um
oropharynx or'-ō-FAR-inks
osmosis os-MŌ-sis
oxaloacetic ok'-sa-lō-a-SĒ-tik
oxidative OK-si-dā-tiv
oxyntic ok-SIN-tik

palate PAL-at
palatoglossal pal'-a-tō-GLOS-al
palatopharyngeal pal'-a-tō-fa-RIN-jē-al
pancreas PAN-krē-as
pancreatic pan'-krē-AT-ik
Paneth PA-nāt
pantothenic pan'-tō-THĒ-nik
papilla pa-PIL-a
papillae pa-PIL-ē
parietal pa-RĪ-e-tal
parotid pa-ROT-id
pentose PEN-tōs
pepsin PEP-sin
pepsinogen pep-SIN-ō-jen
peptidase PEP-ti-dās
peptide PEP-tīd
periodontal per'-ē-ō-DON-tal
periosteum per'-ē-OS-tē-um
peristalsis per'-i-STAL-sis
peritoneal per'-i-TŌN-ē-al
peritoneum per'-i-tō-NĒ-um
Peyer's PĪ-erz
phagocyte FAG-ō-sīt
phagocytic fag'-ō-SIT-ik
phagocytosis fag'-ō-sī-TŌ-sis
pharyngeal fa-RIN-jē-al
pharynx FAR-inks
phosphatase FOS-fa-tās'
phosphorus FOS-fō-rus
phosphorylation fos'-for-i-LĀ-shun
plaque PLAK
plasma PLAZ-ma
plexus PLEK-sus
plicae circulares PLĪ-kē SER-kyoo-lar-es
polypeptide pol'-ē-PEP-tīd
polysaccharide pol'-ē-SAK-a-rīd
polyunsaturated pol'-ē-un-SACH-e-rā-ted

preoptic   prē-OP-tik
procarboxypeptidase   prō'-kar-bok'-sē-PEP-ti-dās
proctology   prok-TOL-ō-jē
prostaglandin   pros'-ta-GLAN-din
protein   PRO-tēn
pyloric   pī-LOR-ik
pylorus   pī-LOR-us
pyridoxine   pēr'-i-DOK-sēn

rectum   REK-tum
regurgitate   rē-GUR-ji-tāt'
reticuloendothelial   re-tik'-yoo-lō-en'-dō-THĒ-lē-al
retinal   RE-ti-nal
retroperitoneal   re'-trō-per-i-tō-NĒ-al
riboflavin   rī'-bō-FLĀ-vin
ribonuclease   rī-bō-NOO-klē-ās
ribonucleic   rī-bō-noo-KLĀ-ik
Rivinus   rē-VĒ-nus
rugae   ROO-jē

saliva   sa-LĪ-va
salivary   SAL-i-ver-ē
salivation   sal-i-VĀ-shun
salivatory   SAL-i-va-tor'-ē
Santorini   san'-tō-RĒ-nē
satiety   sa-TĪ-e-tē
secretin   se-KRĒ-tin
serosa   se-RŌ-sa
serous   SĒR-us
sigmoid   SIG-moyd
sinusoid   SĪN-yoo-soyd
sinusoidal   SĪN-yoo-soyd-al
somatostatin   sō'-ma-tō-STAT-in
sphincter   SFINGK-ter
splenic   SPLEN-ik
stellate   STEL-āt
Stensen's   STEN-senz
steroid   STER-oyd
sublingual   sub-LING-gwal
submandibular   sub'-man-DIB-yoo-lar
submaxillary   sub'-MAK-si-lar-ē
submucosa   sub-myoo-KŌ-sa
submucosal   sub-myoo-KŌ-sal
substrate   SUB-strāt
sucrase   SOO-krās
sucrose   SOO-krōs

taeniae coli   TĒ-nē-a KŌ-lī
testosterone   tes-TOS-te-rōn
thiamine   THĪ-a-min
thyroxine   thī-ROK-sēn
tocopherol   tō-KOF-er-ol
transamination   trans'-am-i-NĀ-shun
triacylglycerol   trī-as'-il-GLIS-er-ol

triglyceride   trī-GLIS-er-īd
trypsin   TRIP-sin
trypsinogen   trip-SIN-ō-jen

uvula   YOO-vyoo-la

Vater   VA-ter
vermiform   VER-mi-form
vermilion   ver-MIL-yon
vestibule   VES-ti-byool
villi   VIL-ī
villus   VIL-lus
viscera   VIS-er-a
visceral   VIS-er-al
viscus   VIS-kus

Wharton's   HWAR-tunz
Wirsung   VĒR-sung

zymogenic   zī-mō-JEN-ik

Glossary of Terms

Absorption   The passage of digested foods from the gastrointestinal tract into blood or lymph.

Absorptive state   Metabolic state during which ingested nutrients are being absorbed by the blood or lymph from the gastrointestinal tract. Also called the *fed state*.

Accessory duct (of pancreas)   A duct of the pancreas that empties into the duodenum about 2.5 (1 in.) superior to the hepatopancreatic ampulla (ampulla of Vater). Also called the *duct of Santorini*.

Accessory structures (of GI tract)   Teeth, tongue, salivary glands, liver, gallbladder, and pancreas.

Acetaldehyde   A product of alcohol catabolism (breakdown) that damages the mitochondria in liver cells. (Alcohol catabolism occurs in liver cells.)

Acetyl coenzyme A (Actyl CoA)   An acetyl group (2-carbon fragment) attached to coenzyme A. In this form an acetyl group is carried to the Krebs cycle, where it combines with a molecule of oxaloacetic acid, forming citric acid.

Acetyl group   A 2-carbon fragment produced when pyruvic acid loses a molecule of carbon dioxide.

Acid   A proton donor, or substance that dissociates into hydrogen ions (H^+) and anions; characterized by an excess of hydrogen ions and a pH less than 7.

Acinar cell   A cell in the pancreas that secretes pancreatic juice (fluid and digestive enzymes).

Acini   Masses of acinar cells in the pancreas that secrete digestive enzymes.

Active transport   The movement of substances across cell membranes against a concentration gradient, requiring the expenditure of energy (ATP).

Adenosine diphosphate (ADP)   When a phosphate is split from ATP, the resulting molecule contains two phosphate groups and is called ADP. The energy released during this reaction is used for cell functions.

Adenosine triphosphate (ATP)   The universal energy-carrying molecule manufactured in all living cells. It consists of adenine (a nitrogenous base), ribose (a 5-carbon sugar), and three phosphate groups.

Adipose tissue   Tissue composed of adipocytes (fat cells) specialized for triglyceride storage and present in the form of soft pads between various organs for support, protection, and insulation.

ADP   *See* Adenosine diphosphate.

Adrenaline   *See* Epinephrine.

Adventitia   The outermost covering of a structure or organ.

Aggregated lymphatic follicles   Aggregated lymph nodules that are most numerous in the ileum (of the small intestine). Also called *Peyer's patches*.

Alimentary canal   *See* Gastrointestinal tract.

Alpha cell   An endocrine cell in a pancreatic islet (islet of Langerhans) that secretes the hormone glucagon.

Alpha dextrinase   A brush border enzyme (an enzyme present in the cell membrane of the microvilli in the the small intestine). It splits off one glucose unit at a time from alpha-dextrin molecules.

Alveolar process   An arch of the mandible or maxilla that contains the tooth sockets (alveoli).

Alveolus (plural: alveoli)   Tooth socket in the alveolar process of the mandible (lower jaw) or maxilla (upper jaw).

Amino acid   An organic acid that is the building unit from which proteins are formed. There are twenty different kinds of amino acids. Linked in different sequences by chemical bonds, they form different proteins.

Amino group $-NH_2$.

Aminopeptidase   A brush border enzyme. It splits off the terminal amino acid from the amino end of a peptide.

Ampulla   A saclike dilation of a canal.

Ampulla of Vater   *See* Hepatopancreatic ampulla.

Amylase   *See* Pancreatic amylase and Salivary amylase.

Anabolism   Synthesis. Energy-requiring reactions in which small molecules are built up into larger ones.

Anal canal   The terminal 2 or 3 cm (1 in.) of the rectum.

Anal column   A longitudinal fold in the mucous membrane of the anal canal; contains a network of arteries and veins.

Anal triangle   The subdivision of the female or male perineum that contains the anus.

Anion   A negatively charged ion.

Antrum (of stomach)   The wide portion of the pylorus that connects to the body of the stomach. Also called the *pyloric antrum*.

Anus   The distal end and outlet of the rectum.

Apical foramen   An opening at the base of each root canal through which blood vessels, lymphatic vessels, and nerves enter and exit the tooth.

Apoferritin   A protein found in liver cells which combines with iron to form ferritin, the form in which iron is stored in the liver.

Apoprotein   Protein carrier. Apoproteins combine with lipids to form lipoproteins, which are soluble in blood plasma.

Appendix   *See* Vermiform appendix.

Areolar connective tissue   A type of loose connective tissue. The lamina propria (the middle layer of the mucosa) consists of areolar connective tissue.

Ascending colon   The portion of the large intestine that passes upward from the cecum to the lower edge of the liver where it bends at the right colic (hepatic) flexure to become the transverse colon.

Ascorbic acid   *See* Vitamin C.

ATP   *See* Adenosine triphosphate.

Autonomic plexus   An extensive network of sympathetic and parasympathetic nerve fibers.

Baby teeth   *See* Deciduous teeth.

Bacteria (singular : bacterium)   All prokaryotic organisms, except the blue-green algae. Prokaryotic organisms are single-celled organisms that do not contain a membrane-bound nucleus. (*pro* = before; *karyon* = nucleus)

Barbiturate   A sedative. A derivative of barbituric acid.

Basal metabolic rate (BMR)   The rate of metabolism measured under standard or basal conditions.

Base   A proton acceptor, or a nonacid; characterized by an excess of hydroxide ions (OH^-) and a pH greater than 7.

Basement membrane   A thin, extracellular layer consisting of the basal lamina (secreted by epithelial cells) and the reticular lamina (secreted by connective tissue cells).

Beta-carotene   A source of vitamin A. Widely distributed in green leafy vegetables and yellow vegetables. Also called *provitamin A*.

Beta cell   An endocrine cell in a pancreatic islet (islet of Langerhans) that secretes the hormone insulin.

Beta oxidation   The conversion of fatty acids into acetyl coenzyme A (acetyl CoA). Acetyl CoA enters the Krebs cycle and is used for the production of energy (ATP).

Bicarbonate ion HCO_3^-.

Bicuspid   A tooth with 2 cusps and 1 root. Used to crush and grind food. 2 pairs in each jaw. Also called a *premolar*.

Bile   A secretion of the liver that is stored in the gallbladder. Consists of water, bile salts, bile pigments, cholesterol, lecithin, and several ions. It causes the dispersion of large fat droplets into smaller fat droplets (emulsification).

Bile capillary   *See* Canaliculus.

Bile ducts   Ducts that carry bile. Includes the canaliculi (bile capillaries), hepatic ducts, bile ducts, and hepatopancreatic ampulla.

Biliary   Relating to bile, the gallbladder, or the the bile ducts.

Bilirubin   A red pigment that is one of the end products of hemoglobin breakdown in the liver cells; it is excreted as a waste material in the bile.

Biliverdin   A green pigment that is one of the first products of hemoglobin breakdown in liver cells; it is converted to bilirubin or excreted as a waste material in bile.

Biotin   One of the B vitamins. Functions as a coenzyme involved in fat synthesis, amino acid metabolism, and glycogen formation.

BMR   *See* Basal metabolic rate.

Body   (of stomach)   The large central region of the stomach below the fundus.

Bolus   A soft, rounded mass, usually food, that is swallowed.

Brunner's gland   *See* Duodenal gland.

Brush border   The fuzzy line that is seen in a photomicrograph of the intestinal lining (taken through a light microscope). It is caused by the microvilli (fingerlike projections) on the apical (free) surface of the absorptive epithelial cells.

Brush border enzymes   Digestive enzymes embedded in the plasma membrane of the microvilli of epithelial cells lining the small intestine. Digestion by these enzymes occurs at the surface of the epithelial cells, rather than in the lumen of the small intestine. Carbohydrate-digesting enzymes are alpha-dextrinase, maltase, sucrase, and lactase; protein-digesting enzymes include aminopeptidase and dipeptidase; nucleotide-digesting enzymes are nucelosidases and phosphatases.

Buccal   Pertaining to the cheek or mouth.

Buffer system   A pair of chemicals, one a weak acid and one the salt of the weak acid, which functions as a weak base. Buffer systems resist changes in the pH.

Bulk   *See* Fiber.

Bulk minerals   Minerals required in amounts in excess of 100 mg/day. Include calcium (Ca), chlorine (Cl), magnesium (Mg), phosphorus (P), potassium (K), sodium (Na), and sulfur (S).

B Vitamins B_1 (thiamine); B_2 (riboflavin); niacin (nicotinic acid); B_6 (pyridoxine); folacin (folic acid); B_{12} (cobalamin); biotin; and pantothenic acid.

Calcium   A bulk mineral. Important for bone development, nerve and muscle function, and blood clotting.

Calorie   A unit of heat. A calorie (cal) is the amount of heat necessary to raise 1 g of water 1°C from 14° to 15°C. The kilocalorie (kcal) is 1,000 calories. In nutrition studies, the calories in food are technically kilocalories.

Canaliculus   (plural: canaliculi)   A small channel or canal. In the liver, also called a *bile capillary*.

Canine   *See* Cuspid.

Carbohydrate   An organic compound containing carbon, hydrogen, and oxygen in a particular amount and arrangement. It consists of sugar units and has the general formula $(CH_2O)n$. *Also see* Complex carbohydrates and Simple carbohydrates.

Carboxyl group   –COOH.

Carboxypeptidase   A digestive enzyme found in pancreatic juice; activated from procarboxypeptidase by trypsin. It splits off the terminal amino acid from the carboxyl end of a peptide.

Cardia   A narrow band about 2 cm in width that surrounds the opening between the stomach and the esophagus.

Catabolism   Chemical reactions that break down complex organic compounds into simple ones with the release of energy.

Cation   A positive ion.

Caudate   *See* Lobes of the liver.

CCK   *See* Cholecystokinin.

Cecum   A blind pouch at the proximal end of the large intestine to which the ileum is attached.

Cellular respiration   The oxidation of glucose in the cells of the body. Glucose combines with oxygen to form carbon dioxide and water; energy in the form of ATP is released during this series of reactions.

Cellulose   An indigestible carbohydrate (polysaccharide). Cellulose fibers surround the cell membranes of plant cells, forming the cell walls.

Cementum   Calcified tissue covering the root of a tooth.

Central vein   A vein that passes through the center of a liver lobule.

Cephalic   Pertaining to the head; superior in position.

Cephalic phase   (of gastric digestion)   Reflexes initiated by sensory receptors in the head. Parasympathetic impulses travel from the medulla via the vagus nerve to the stomach, promoting peristalsis and stimulating the secretion of gastric juice.

Cheeks   The lateral walls of the mouth.

Chemical reaction   The combination or breaking apart of atoms in which chemical bonds are formed or broken and new products with different properties are produced.

Chief cell   A cell in the mucous membrane lining of the stomach that secretes pepsinogen (the precursor of the enzyme pepsin). Also called a *zymogenic cell*.

Chlorine   A bulk mineral. Important for acid-base balance, water balance, and the formation hydrochloric acid (HCl).

Cholecystokinin (CCK)   A hormone secreted by enteroendocrine cells in the lining of the small intestine. It stimulates the secretion of pancreatic juice that is rich in digestive enzymes and causes the ejection of bile from the gallbladder.

Cholesterol   A steroid (type of lipid). A component of cell membranes; used for the synthesis of steroid hormones and bile salts.

Chylomicron   A protein-coated spherical structure that contains triglycerides, phospholipids, and cholesterol. It is absorbed from the gastrointestinal tract into a lymphatic capillary (lacteal) and carried to the blood via lymphatic vessels.

Chyme The semifluid mixture of partly digested food and digestive secretions. Found in the stomach and small intestine during the digestion of a meal.

Chymotrypsin A digestive enzyme found in pancreatic juice; it is activated from chymotrypsinogen by trypsin. It breaks down proteins into smaller fragments called peptides.

Circular folds Permanent, deep transverse folds in the mucosa and submucosa of the small intestine that increase the surface area for absorption. Also called *plicae circulares*.

Circular muscle The inner muscle layer of the muscularis. (The muscularis is the layer of the GI tract wall between the serosa and the submucosa.)

Circumvallate papilla One of the circular projections located in the posterior portion of the tongue; contains taste buds.

Citric acid cycle *See* Krebs cycle.

Cobalamin *See* Vitamin B$_{12}$.

Coenzyme A nonprotein organic molecule that generally serves as a carrier that transfers atoms or small molecular fragments from one reaction to another. An example is NAD, which transfers hydrogen atoms from the Krebs cycle to the electron transport chain.

Coenzyme A (CoA) A coenzyme derived from pantothenic acid, a B vitamin. It serves as a carrier molecule to guide the 2-carbon acetyl group into the Krebs cycle.

Colic Relating to the colon.

Colon The division of the large intestine consisting of ascending, transverse, descending, and sigmoid portions.

Common bile duct A tube formed by the union of the common hepatic duct (from the liver) and the cystic duct (from the gallbladder). It empties bile into the duodenum at the hepatopancreatic ampulla (ampulla of Vater).

Common hepatic duct A duct formed by the union of the left and right hepatic ducts. It joins the cystic duct (from the gallbladder) to form the common bile duct.

Complementary proteins Combinations of plant proteins that include all the essential amino acids. An example is rice and beans.

Complex carbohydrates Starches (polysaccharides). Found in many foods including cereals, breads, dry beans, peas, and potatoes.

Complex protein A protein that contains all of the essential amino acids. Usually of animal origin.

Core temperature The body's temperature below the skin and subcutaneous tissue. Normally about 37 °C (98.6 °F).

Cortisol A hormone secreted by the adrenal cortex. It stimulates gluconeogenesis, lipolysis, and protein synthesis.

Crown The exposed portion of a tooth above the level of the gums.

Crypt of Lieberkühn *See* Intestinal gland.

Cusp A pointed surface on the crown of a tooth.

Cuspid A tooth with 1 cusp and 1 root. 1 pair in each jaw. Also called a *canine*.

Cystic duct The duct that transports bile from the gallbladder to the common bile duct.

Cytochrome enzyme system *See* Electron transport chain.

Deamination Removal of an amino group (–NH$_2$) from an amino acid, forming a keto acid (a type of carbohydrate).

Deciduous teeth The first set of 20 teeth, which are lost during childhood. Also called *milk teeth* or *baby teeth*.

Defecation The discharge of feces from the rectum.

Deglutition Swallowing.

Delta cell An endocrine cell in a pancreatic islet (islet of Langerhans) that secretes growth hormone-inhibiting hormone (somatostatin).

Dentes Teeth.

Dentin The bony tissues of a tooth enclosing the pulp cavity.

Dentition The eruption of teeth. The number, shape, and arrangement of teeth.

Deoxyribonuclease A digestive enzyme found in pancreatic juice. It splits deoxyribonucleic acid (DNA) into nucleotides.

Deoxyribonucleic acid (DNA) A double-stranded nucleic acid in the shape of a double helix. Consists of repeating units called nucleotides. The nucleotides are composed of the sugar deoxyribose, a phosphate group, and one of four nitrogenous bases (adenine, cytosine, guanine, or thymine). Genetic information is encoded in the sequence of the nucleotides.

Descending colon The part of the large intestine descending from the left colic (splenic) flexure to the level of the left iliac crest.

Diffusion A passive process in which there is a net movement of molecules or ions from a region of high concentration to a region of low concentration until equilibrium is reached.

Digestion The mechanical and chemical breakdown of food into small molecules that can be absorbed and used by body cells.

Dipeptidase A brush border enzyme. It splits dipeptides (two amino acids joined by a peptide bond) into amino acids.

Dipeptide Two amino acids linked by a peptide bond.

Disaccharide A double sugar. Examples are maltose (2 glucose units), sucrose (glucose and fructose), and lactose (glucose and galactose).

Diverticulum A sac or pouch in the wall of a canal or organ, especially in the colon.

DNA *See* Deoxyribonucleic acid.

Duct of Rivinus *See* Sublingual gland.

Duct of Santorini *See* Accessory duct.

Duct of Wirsung *See* Pancreatic duct.

Duodenal gland Gland in the submucosa of the duodenum that secretes an alkaline mucus. The mucus protects the lining of the small intestine from the action of digestive enzymes and helps to neutralize the acidic contents (chyme) of the duodenum. Also called *Brunner's gland*.

Duodenal papilla An elevation on the duodenal mucosa that receives the hepatopancreatic ampulla (ampulla of Vater).

Duodenum The first 25 cm (10 in.) of the small intestine.

Electrolyte Any compound that separates into ions when dissolved in water.

Electron transport chain A chain of carrier molecules (enzymes) located on the inner membrane of a mitochondrion. As electrons are passed along the chain, energy is released for the production of ATP. Also called the *electron transport system* or *cytochrome enzyme system*.

Electron transport system *See* Electron transport chain.

Emulsification The coating of small lipid droplets with bile salts. This prevents the small droplets from combining to form larger droplets.

Enamel The hard, white substance covering the crown of a tooth.

Endocrine gland A gland that secretes hormones into the blood; a ductless gland.

Endodontics The branch of dentistry concerned with the prevention, diagnosis, and treatment of diseases that affect the pulp, root, periodontal ligament, and alveolar bone.

Endopeptidase A protein-digesting enzyme that acts on the interior peptide bonds of a protein molecule. Examples are trypsin and chymotrypsin.

Energy The capacity to do work.

Enteric Pertaining to the small intestine.

Enteric nervous system A network of autonomic nerve fibers located in the wall of the GI tract, which controls motility (muscular contractions) of the GI tract.

Enteroendocrine cells Cells found among the epithelial cells lining the GI tract. Secrete digestive hormones (gastrin, cholecystokinin (CCK), gastric inhibitory peptide (GIP), and secretin).

Enterogastric reflex A neural reflex initiated by the presence of food in the small intestine. Nerve impulses carried to the medulla (in the brain stem) from the duodenum return to the stomach and inhibit gastric secretion and motility.

Enterokinase An activating enzyme secreted by cells in the the mucous lining of the small intestine. Converts the inactive enzyme trypsinogen into trypsin.

Enzyme A substance that affects the speed of a specific chemical reaction; an organic catalyst, usually a protein.

Epiglottis A large, leaf-shaped piece of cartilage lying on top of the the larynx, with its "stem" attached to the thyroid cartilage and its "leaf" portion unattached and free to move up and down to cover the glottis (vocal folds and rima glottidis).

Epinephrine A hormone secreted by the adrenal medulla. It produces actions similar to those that result from sympathetic nerve stimulation. Also called *adrenaline*.

Epiploic appendages Small fat-filled pouches of visceral peritoneum attached to the large intestine.

Epithelial tissue The tissue that forms glands or the outer part of the skin and lines blood vessels, hollow organs, and passages that lead externally from the body.

Epithelium Epithelial tissue. Covering and lining epithelium forms the outer layer of the skin and the outer layer of some internal organs; it forms the inner lining of blood vessels, ducts, body cavities, and the interiors of the respiratory, digestive, urinary, and reproductive systems. Glandular epithelium constitutes the secreting portion of glands, such as sweat glands and the thyroid gland.

Esophageal hiatus The opening in the diaphragm through which the esophagus passes.

Esophageal stage (of swallowing) The third stage of swallowing, when a bolus (mass of food) passes through the esophagus into the stomach. An involuntary stage.

Esophagus A hollow muscular tube connecting the pharynx and the stomach.

Essential amino acids Amino acids that cannot be synthesized by the body; they must be included in the diet.

Essential fatty acids Fatty acids that cannot be synthesized by the body; they must be included in the diet.

Excrement *See* Feces.

Excretion The process of eliminating waste products from a cell, tissue, or the entire body; or the products excreted.

Exocrine gland A cluster of epithelial cells that secretes substances (such as oil, sweat, mucus, or digestive en-

zymes) into ducts that lead to an epithelial surface.

Exopeptidase A protein-digesting enzyme that splits amino acids from the end of a protein molecule. Examples are carboxypeptidase and aminopeptidase.

External anal sphincter A ring of skeletal (voluntary) muscle in the anus that functions as a valve, controlling defecation.

Extrinsic muscles (of tongue) Skeletal muscles that originate outside the tongue and insert into it.

Facial nerve 7th cranial nerve. Innervates the lacrimal gland, sublingual gland, submandibular gland, and the anterior two-thirds of the tongue.

Facilitated diffusion Diffusion in which a substance not soluble by itself in lipids is transported across a selectively permeable membrane by combining with a transporter (carrier molecule).

Falciform ligament A sheet of parietal peritoneum between the two principal lobes of the liver.

Fasting state *See* Postabsorptive state.

Fat *See* Lipid.

Fat-soluble vitamins Vitamins A, D, E, and K.

Fatty acid A hydrocarbon chain with a carboxyl group at one end. The hydrocarbon chain is nonpolar (insoluble in water); the carboxyl group (–COOH) can function as an acid by releasing a hydrogen ion. Three fatty acids linked to a molecule of glycerol form a triglyceride (neutral fat).

Fauces The opening from the mouth into the pharynx.

F cell A cell in a pancreatic islet (islet of Langerhans) that secretes the hormone pancreatic polypeptide.

Fecal matter *See* Feces.

Feces Material discharged from the rectum and made up of bacteria, excretions, and food residue. Also called *fecal matter* or *excrement* or *excretion* or *stool*.

Fed state *See* Absorptive state.

Ferritin The form in which iron is stored in the liver. Liver cells contain a protein called apoferritin, which combines with iron to form ferritin.

Fiber Indigestible complex carbohydrates (starches) such as cellulose and pectin. Large quantities are found in bran, fruits, whole-grain cereals, whole-grain breads, and vegetables. Also called *bulk* or *roughage*.

Flatus Air (gas) in the stomach or intestines. Commonly used to denote passage of gas rectally.

Flexure A bend, curve, or turn.

Folacin One of the B vitamins. Functions as a coenzyme involved in amino acid and nucleic acid metabolism. Also called *folic acid*.

Folic acid *See* Folacin.

Foramen An opening or hole.

Frenulum A small fold of mucous membrane that connects two parts and limits movement.

Fructose A monosaccharide (single sugar). The disaccharide sucrose (table sugar) consists of fructose and glucose.

Fundus (of stomach) The rounded portion of the stomach located above and to the left of the cardia (region adjacent to the opening to the esophagus).

Galactose A monosaccharide (single sugar). The disaccharide lactose (milk sugar) consists of galactose and glucose.

Gallbladder A small pouch that stores bile. Located under the liver.

Gastric emptying   Emptying of the stomach. The hormone gastrin causes contraction of the lower esophageal sphincter, relaxation of the pyloric sphincter, and increased contractions of the stomach wall.

Gastric gland   A gland in the stomach lining at the bottom of a gastric pit.

Gastric inhibitory peptide (GIP)   A hormone secreted by enteroendocrine cells located in the lining of the small intestine. It inhibits the secretion of gastric juice and gastric emptying.

Gastric juice   The secretions of three types of cells found in the lining of the stomach: mucous cells (secrete mucus); parietal cells (secrete hydrochloric acid); and chief cells (secrete pepsinogen).

Gastric phase (of gastric digestion)   Reflexes initiated by sensory receptors in the stomach. They stimulate stomach motility (churning) and the secretion of gastric juice.

Gastric pit   A narrow channel formed by the lining of the stomach. It extends down into the lamina propria (connective tissue layer) of the mucosa (mucous membrane).

Gastrin   A hormone secreted by G cells in the lining of the stomach. It stimulates the secretion of gastric juice and gastric motility.

Gastroenterology   The medical specialty that deals with the structure, function, diagnosis, and treatment of diseases of the stomach and intestines.

Gastrointestinal tract   A continuous tube running through the ventral body cavity extending from the mouth to the anus. Also called the *GI tract* or the *alimentary canal*.

G cell   Cell located in the mucosa of the pylorus (of the stomach). Secretes the hormone gastrin into the blood.

GHIH   *See* Growth hormone-inhibiting hormone.

Gingivae (singular: gingiva)   Gums.

GIP   *See* Gastric inhibitory peptide.

GI tract   *See* Gastrointestinal tract.

Gland   Single or group of specialized epithelial cells that secrete substances.

Glossopharyngeal nerve   9th cranial nerve. Innervates the parotid gland (salivary gland), the posterior one-third of the tongue, the throat, and the carotid sinus (of the carotid artery).

Glottis   The vocal folds (true vocal cords) in the larynx plus the space between them (rima glottidis).

Glucagon   A hormone. Produced by the alpha cells of the pancreas. Increases the blood glucose level.

Gluconeogenesis   The conversion of a substance other than carbohydrate into glucose.

Glucose   A six-carbon sugar, $C_6H_{12}O_6$. The major energy source for every cell type in the body.

Glucose 6-phosphate   An intermediate metabolite in the glycolytic pathway (conversion of glucose into pyruvic acid).

Glucose-sparing reactions   Reactions that occur during the postabsorptive state (fasting state). All body cells (except neurons) reduce their oxidation of glucose and switch over to fatty acids as their main source of ATP.

Glyceraldehyde 3-phosphate   An intermediate metabolite in the glycolytic pathway (conversion of glucose into pyruvic acid).

Glycerol   A 3-carbon molecule, which is a component of a triglyceride molecule. A triglyceride (neutral fat) consists of three fatty acids chemically linked to a molecule of glycerol.

Glycogen   Molecule used for the storage of glucose in the liver and skeletal muscles. A highly branched polymer of glucose containing thousands of glucose molecules (subunits).

Glycogenesis   Glycogen synthesis. The process by which many molecules of glucose combine to form glycogen.

Glycogenolysis   Breakdown of glycogen. The process of converting glycogen to glucose.

Glycolysis   Series of chemical reactions in which a molecule of glucose is split into two molecules of pyruvic acid. Occurs in the cytosol of a cell.

Goblet cell   A goblet-shaped unicellular gland that secretes mucus. Also called *mucous cell*.

Greater curvature (of stomach)   The convex lateral border of the stomach.

Greater omentum   A large fold in the serosa (outer covering) of the stomach that hangs down like an apron over the front of the intestines.

Growth hormone   *See* Human growth hormone.

Growth hormone-inhibiting hormone (GHIH)   A hormone secreted by neurons in the hypothalamus and by D cells in the pancreatic islets. It inhibits the release of human growth hormone from the anterior pituitary gland. Also called *somatostatin*.

Hard palate   The anterior portion of the roof of the mouth. Formed by the maxillae and palatine bones and lined by mucous membrane.

Haustra (singular: haustrum)   A series of pouches, which give the colon its puckered appearance.

Haustral churning   Contractions of the haustra. When stretched to a certain degree, haustra contract and squeeze their contents into the next portion of the colon.

HCl   *See* Hydrochloric acid.

HDL   *See* High-density lipoprotein.

Heat   A form of kinetic energy. Can be measured as temperature and is expressed in units called calories.

Heat-losing center   A region in the hypothalamus that controls the responses that lower body temperature.

Heat-promoting center   A region in the hypothalamus that controls the responses that raise body temperature.

Hepatic   Refers to the liver.

Hepatic cells   Specialized epithelial cells arranged in irregular, branching, interconnected plates around a central vein in the liver. Also called *hepatocytes* or *liver cells*.

Hepatic duct   A duct that receives bile from the bile capillaries. Small hepatic ducts merge to form the larger right and left hepatic ducts that unite to leave the liver as the common hepatic duct.

Hepatic flexure   *See* Right colic flexure.

Hepatic portal circulation   The flow of blood from the stomach, spleen, pancreas, and intestines to the liver.

Hepatocytes   *See* Hepatic cells.

Hepatopancreatic ampulla   The dilated duct formed by the union of the common bile duct and the pancreatic duct; it empties into the duodenum. Also called the *ampulla of Vater*.

High-density lipoproteins (HDLs)   "Good cholesterol." Lipoproteins that remove excess cholesterol from body cells and transport it to the liver for elimination.

Hiatus   An opening; a foramen.

Human growth hormone (hGH)   A hormone secreted by cells in the anterior pituitary gland. It brings about growth of body tissues, especially skeletal and muscular tissues. Also called *growth hormone* and *somatotropin*.

Hydrochloric acid (HCl)   An acid secreted by the parietal (oxyntic) cells of the stomach lining. Kills microbes in food; denatures proteins; converts pepsinogen into pepsin, which digests proteins into peptides; inhibits secretion of the hormone gastrin; and stimulates secretion of the hormones secretin and cholecystokinin (CCK).

Hydrolysis (*hydro* = water; *lysis* = to split)   To split apart by using water. Large molecules such as carbohydrates, lipids, and proteins are broken into smaller molecules by hydrolysis.

Hydrolytic reaction   A reaction that involves hydrolysis.

Hypopharynx   *See* Laryngopharynx.

Ileocecal sphincter   A fold of mucous membrane that guards the opening from the ileum into the large intestine. Also called the *ileocecal valve*.

Ileocecal valve   *See* Ileocecal sphincter.

Ileum   The terminal portion of the small intestine.

Iliac crest   The superior border of the ilium.

Incisor   A chisel-shaped tooth with 1 root. 2 pairs in each jaw (central and lateral incisors).

Incomplete protein   A protein that lacks one or more of the essential amino acids.

Ingestion   The taking in of food, liquid, or drugs by mouth.

Inorganic compound   One of the two main types of chemical compounds. Inorganic compounds usually lack carbon; usually are small and contain ionic bonds. Examples include water and many acids, bases, and salts.

Insulin   A hormone secreted by beta cells of the pancreatic islets. Decreases the blood glucose level.

Intermediate metabolite   In a series of reactions (metabolic pathway), the substances (metabolites) that are not final products.

Internal anal sphincter   A ring of involuntary smooth muscle in the anus. Helps control defecation.

Intestinal gland   A gland that opens onto the surface of the intestinal mucosa and secretes digestive enzymes. Also called a *crypt of Lieberkühn*.

Intestinal phase   (of gastric digestion)   Reflexes initiated by sensory receptors in the small intestine. They inhibit stomach motility (churning) and the secretion of gastric juice.

Intrinsic factor   A glycoprotein formed by cells in the lining of the stomach that is necessary for the absorption of vitamin B_{12}.

Intrinsic muscles (of tongue)   Skeletal muscles that originate and insert within the tongue.

Islet of Langerhans   *See* Pancreatic islet.

Jejunum   The middle portion of the small intestine.

Keto acid   A type of carbohydrate. During the absorptive state (fed state) many amino acids that enter liver cells are deaminated (removal of an amino group), forming keto acids.

Ketogenesis   The production of ketone bodies. During excessive triglyceride catabolism, liver cells can combine two acetyl coenzyme A molecules to form a ketone body (acetoacetic acid).

Ketones   *See* Ketone bodies.

Ketone bodies   Substances (acetone, acetoacetic acid, and beta-hydroxybutyric acid) produced during excessive triglyceride catabolism. Also called *ketones*.

Kilocalorie (kcal)   The unit of heat used to express the heating value of foods and to measure metabolic rate. 1 kilocalorie = 1000 calories.

Krebs cycle   A series of energy-yielding chemical reactions that occur in the matrix of mitochondria in which energy is transferred to carrier molecules for subsequent liberation and carbon dioxide is formed. Also called *citric acid cycle* or *tricarboxylic acid cycle (TCA cycle)*.

Kupffer's cell   *See* Stellate reticuloendothelial cell.

Labia   (singular: labium)   Lips.

Labial frenulum   A medial fold of mucous membrane between the inner surface of the lip and the gums.

Lactase   A brush border enzyme. It breaks down the disaccharide lactose into galactose and glucose.

Lacteals   Lymphatic capillaries in the villi of the small intestine. They absorb protein-coated lipid droplets called chylomicrons that are formed in the epithelial cells lining the small intestine.

Lactose   A disaccharide (double sugar) consisting of galactose and glucose.

Lamina propria   The connective tissue layer of a mucous membrane. Located between the epithelium (which lines the gastrointestinal tract) and the muscularis mucosae.

Large Intestine   The portion of the gastrointestinal tract extending from the ileum of the small intestine to the anus. It is divided into the cecum, colon, rectum, and anal canal.

Laryngopharynx   The inferior portion of the pharynx, extending downward from the level of the hyoid bone to divide posteriorly into the esophagus and anteriorly into the larynx. Also called the *hypopharynx*.

Larynx   A short passageway that connects the pharynx (throat) and the trachea (windpipe). Contains the vocal cords. Also called the *voice box*.

LDL   *See* Low-density lipoprotein.

Left colic flexure   The portion of the transverse colon that forms a bend (flexure) under the spleen and leads into the descending colon. Also called the *splenic flexure*.

Lesser curvature (of stomach)   The concave medial border of the stomach.

Lesser omentum   A fold of the peritoneum that extends from the liver to the lesser curvature of the stomach and the duodenum.

Ligamentum teres (of liver)   *See* Round ligament.

Lingual frenulum   A fold of mucous membrane that connects the tongue to the floor of the mouth.

Lingual lipase   A digestive enzyme secreted by glands on the dorsum of the tongue. Splits a triglyceride molecule, forming one monoglyceride and two fatty acids.

Lipase   *See* Lingual lipase and Pancreatic lipase.

Lipid   An organic compound composed of carbon, hydrogen, and oxygen that is usually insoluble in water, but soluble in alcohol, ether, and chloroform. Examples include triglycerides, phospholipids, steroids, and prostaglandins. Also called *fat*.

Lipid profile   Blood test that measures total cholesterol, high-density lipoprotein, low-density lipoprotein, and triglycerides.   Used to assess the risk for cardiovascular disease.

Lipogenesis   The synthesis of lipids from glucose or amino acids by liver cells.

Lipolysis   Triglyceride breakdown. During the postabsorptive state (fasting state), triglycerides in adipose tissue are broken down and fatty acids are released into the blood.

Lipoprotein   A combination of lipids and proteins. Makes lipids water-soluble for transportation in the blood.

Lipoprotein lipase   An enzyme found in capillary endothelial cells that breaks down triglycerides. After a high-fat meal, lipoprotein lipase breaks down triglycerides and other lipoproteins into fatty acids and glycerol. The fatty acids diffuse into liver and fat cells and recombine with glycerol produced by the cells to re-form triglycerides.

Lips   *See* Labia.

Liver   Large gland under the diaphragm that occupies most of the right hypochondriac region and part of the epigastric region. It produces bile salts, heparin, and plasma proteins; converts one nutrient into another; detoxifies substances; stores glycogen, minerals, and vitamins; carries on phagocytosis of blood cells and bacteria; and helps activate vitamin D.

Liver cells   *See* Hepatic cells.

Lobe   A curved or rounded projection.

Lobes of the liver   The liver is divided into two principal lobes, a large right lobe and a smaller left lobe, separated by the falciform ligament. The right lobe is considered by many anatomists to include an inferior quadrate lobe and a posterior caudate lobe.

Lobule   (of liver)   The basic structural and functional unit of the liver. Consists of liver cells (hepatocytes) arranged in irregular, branching, interconnected plates around a central vein.

Longitudal muscle   The outer muscle layer of the muscularis. (The muscularis is the layer of the GI tract wall between the serosa and the submucosa.)

Low-density lipoproteins (LDLs)   "Bad cholesterol." LDLs deliver cholesterol to body cells that need it and deposit cholesterol in and around smooth muscle fibers in arteries, forming fatty plaques.

Lower esophageal sphincter   A sphincter muscle (valve) at the lower end of the esophagus.

Lumen   The space within an artery, vein, intestine, or a tube.

Lymphatic capillary   A closed-ended microscopic lymphatic vessel that begins in spaces between cells and converges with other lymphatic capillaries to form lymphatic vessels. Lymphatic capillaries in the villi of the small intestine are called *lacteals*.

Lysozyme   A bactericidal enzyme found in tears, saliva, and perspiration.

Magnesium   A bulk mineral. Important for bone development, nerve and muscle function, and a constituent of coenzymes.

Major duodenal papilla   An elevation of the duodenal mucosa about 10 cm (4 in.) below the pyloric sphincter of the stomach. Location where the hepatopancreatic ampulla empties its contents (bile and pancreatic secretions) into the duodenum.

Maltase   A brush border enzyme. Breaks down the disaccharide maltose into two glucose molecules.

Maltose   A disaccharide (double sugar) consisting of two glucose units.

Mass peristalsis   A strong peristaltic wave that begins at the middle of the transverse colon and drives the contents of the colon into the rectum.

Mastication   Chewing.

Mesentery   A fold of peritoneum attaching the small intestine to the posterior abdominal wall.

Mesoappendix   The mesentery of the appendix. It attaches the appendix to the inferior part of the ileum and adjacent part of the posterior abdominal wall.

Mesocolon   A fold of peritoneum that attaches the small intestine to the posterior abdominal wall.

Mesothelium   The layer of simple squamous epithelium that lines serous cavities.

Metabolism   The sum of all the biochemical reactions that occur within an organism. Includes the synthetic (anabolic) reactions and decomposition (catabolic) reactions.

Metabolite   Any substance produced by metabolism.

Micelle   Small, water-soluble aggregates of lipids (fatty acids and monoglycerides combined with bile salts). It is in this form that fatty acids and monoglycerides reach the epithelial cells of the intestinal villi.

Microvilli (singular: microvillus)   Microscopic, fingerlike projections of the cell membranes of small intestinal cells that increase surface area for absorption.

Milk teeth   *See* Deciduous teeth.

Mineral   Inorganic, homogeneous solid substance that may perform a function vital to life. Examples include calcium, sodium, potassium, iron, phosphorus, and chlorine.

Mixing waves   Peristaltic contractions that pass over the stomach every 20 seconds or so, when food is present in the stomach.

Molar   A tooth with a flattened crown with prominent ridges and 3 roots. Used to crush and grind food. 3 pairs in each jaw (1st molars, 2nd molars, and 3rd molars or wisdom teeth).

Monoglyceride   A product of triglyceride catabolism; a glycerol molecule chemically linked to one fatty acid. The enzyme lipase splits a triglyceride molecule, forming one monoglyceride and two fatty acids.

Monosaccharide   A single sugar. Examples are glucose, fructose, and galactose.

Monounsaturated fatty acid   A fatty acid that contains one double covalent bond between its carbon atoms; it is not completely saturated with hydrogen atoms. Plentiful in triglycerides of olive, peanut, and canola (rapeseed) oils.

Motility   Movement of the contents of the gastrointestinal tract.

Mouth   *See* Oral cavity.

Mucin   A protein found in mucus.

Mucosa   *See* Mucous membrane.

Mucous cell   *See* Goblet cell.

Mucous membrane   A membrane that lines a body cavity that opens to the exterior. Also called the *mucosa*.

Mucus   The thick fluid secretion of mucous glands and mucous membranes.

Muscularis   A muscular layer (coat or tunic) of an organ.

Muscularis mucosae   A thin layer of smooth muscle fibers (cells) located in the outermost layer of the mucosa of the gastrointestinal tract, underlying the lamina propria of the mucosa.

Myenteric plexus   A network of sympathetic and parasympathetic nerve fibers located in the muscularis coat of the small intestine. Also called the *plexus of Auerbach*.

Nasopharynx   The uppermost portion of the pharynx, lying posterior to the nose and extending down to the soft palate.

Neck   A constricted portion of an organ or structure. In a tooth, the constricted region between the crown and the root.

Neutral fat   *See* Triglyceride.

Niacin   One of the B vitamins. A constituent of NAD (the coenzyme used to transport hydrogen atoms from the Krebs cycle to the elctron transport chain). Also called *nicotinic acid*.

Nicotinic acid   *See* Niacin.

Nonessential amino acid   An amino acid that can be synthesized by body cells through transamination (the transfer of an amino group from an amino acid to another substance).

Noradrenaline   *See* Norepinephrine.

Norepinephrine (NE)   A hormone secreted by the adrenal medulla. Produces actions similar to those that result from sympathetic nerve stimulation. Also called *noradrenaline*.

Nuclease   A digestive enzyme found in pancreatic juice that breaks nucleic acids into nucleotides. There are two types: ribonuclease (breaks down RNA) and deoxyribonuclease (breaks down DNA).

Nucleic acids   Ribonucleic acid (RNA) and deoxyribonucleic acid (DNA). An organic compound that is a long polymer of nucleotides, with each nucleotide containing a pentose sugar, a phosphate group, and one of four possible nitrogenous bases (adenine, cytosine, guanine, and thymine or uracil).

Nucleosidase   A brush border enzyme that breaks down nucleotides into nitrogenous bases, pentoses, and phosphates.

Nucleotide   The basic repeating structural unit of a nucleic acid. Each nucleotide consists of three building blocks: a nitrogenous base, a 5-carbon sugar (pentose), and a phosphate group.

Nutrient   A chemical substance in food that provides energy, forms new body components, or assists in the functioning of various body processes.

Oddi's sphincter   *See* Hepatopancreatic ampulla.

Omentum   A fold of the peritoneum that extends from the stomach to adjacent organs.

Oral cavity   The mouth cavity. The space that extends from the gums and teeth to the fauces (opening to the throat). Also called the *mouth*.

Organic compound   Compound that always contains carbon and hydrogen atoms held together by covalent bonds. Examples include carbohydrates, lipids, proteins, and nucleic acids (DNA and RNA).

Oropharynx   The middle portion of the pharynx, lying posterior to the mouth and extending from the soft palate down to the hyoid bone.

Osmosis   The net movement of water molecules through a selectively permeable membrane from an area of high water concentration to an area of lower water concentration until an equilibrium is reached.

Oxaloacetic acid   A 4-carbon intermediate metabolite in the Krebs cycle. It combines with a 2-carbon acetyl group to form the 6-carbon citric acid molecule.

Oxidative phosphorylation   The addition of a phosphate group to a molecule of ADP, forming ATP. Oxygen must be present for this reaction to occur. It occurs in mitochondria in the electron transport chain.

Oxyntic cell   *See* Parietal cell.

Palatoglossal arch   The palatoglossus muscle covered by mucous membrane. Located in the posterior portion of the oral cavity (mouth cavity).

Palatopharyngeal arch   The palatopharyngeus muscle covered by mucous membrane. Located in the posterior portion of the oral cavity (mouth cavity).

Pancreas   A soft, oblong organ lying along the greater curvature of the stomach and connected by a duct to the duodenum. It functions as both an exocrine gland (secreting pancreatic juice via the pancreatic duct) and an endocrine gland (secreting the hormone insulin, glucagon, growth hormone inhibiting hormone, and pancreatic polypeptide).

Pancreatic amylase   A digestive enzyme secreted in pancreatic juice. Splits starches and glycogen (animal starch) into smaller fragments. It does not digest cellulose.

Pancreatic duct   A single, large tube that unites with the common bile duct from the liver and gallbladder and drains pancreatic juice into the duodenum at the hepatopancreatic ampulla (ampulla of Vater). Also called the *duct of Wirsung*.

Pancreatic islet   A cluster of endocrine gland cells in the pancreas. It secretes insulin, glucagon, growth hormone inhibiting hormone, and pancreatic polypeptide. Also called *islet of Langerhans*.

Pancreatic lipase   A digestive enzyme found in pancreatic juice. Splits a triglyceride molecule, forming one monoglyceride and two fatty acids.

Pancreatic polypeptide   A hormone secreted by the F cells of pancreatic islets (islets of Langerhans). It regulates the secretion of pancreatic digestive enzymes.

Paneth cell   A cell in the lining of the small intestine that secretes lysozyme (an enzyme that destroys bacteria).

Pantothenic acid   One of the B vitamins. A constituent of coenzyme A (the coenzyme that carries an acetyl group to the Krebs cycle.)

Papilla   (plural: papillae)   A small nipple-shaped projection or elevation. There are three types of papillae on the tongue: circumvallate, filiform, and fungiform.

Parietal cell   A cell in the lining of the stomach that secretes hydrochloric acid (HCl) and intrinsic factor. Also called *oxyntic cell*.

Parietal peritoneum   A serous membrane that lines the wall of the abdominal cavity.

Parotid gland   One of the paired salivary glands located inferior and anterior to the ears. It is connected to the oral cavity via a duct (Stensen's duct) that opens into the inside of the cheek opposite the upper second molar tooth.

Pentose   A 5-carbon sugar. One of the three building blocks of a nucleotide.

Pepsin   A digestive enzyme. Secreted by chief (zymogenic) cells of the stomach in the inactive form pepsinogen, which is converted to pepsin by hydrochloric acid (secreted by parietal cells). Breaks down proteins into smaller fragments called peptides.

Pepsinogen   The inactive form of pepsin. Secreted by chief (zymogenic) cells of the stomach. Converted to pepsin by hydrochloric acid.

Peptidase   A brush border enzyme that digests proteins. There are two types: aminopeptidase and dipeptidase.

Peptide   A chain of 50 or fewer amino acids linked by peptide bonds.

Peptide bond   A covalent bond linking two amino acids.

Periodontal ligament   A dense fibrous connective tissue that attaches the cementum to the wall of a tooth socket.

Periodontal membrane   The periosteum lining the alveolus (socket) of a tooth.

Periosteum   The membrane that covers bone. Consists of connective tissue, osteoprogenitor cells, and osteoblasts. Essential for bone growth, repair, and nutrition.

Peristalsis   Successive rhythmic contractions along the wall of a hollow muscular structure.

Peritoneal cavity   The potential space between the parietal peritoneum and the visceral peritoneum.

Peritoneal fold   Folds of peritoneum that bind the abdominal organs to each other and to the walls of the abdominal cavity.

Peritoneum   The largest serous membrane of the body. Lines the abdominal cavity and covers the viscera (abdominal organs).

Permanent teeth   A set of 32 teeth. They appear between age 6 and adulthood.

Peyer's patches   *See* Aggregated lymphatic follicles.

Ph   A symbol for the concentration of hydrogen ions in a solution. The pH scale extends from 0 to 14. A pH of 7 indicates neutrality; values less than 7 indicate increasing acidity; values higher than 7 indicate increasing alkalinity.

Phagocyte   A cell that eats bacteria and cellular debris.

Phagocytosis   The process by which cells (phagocytes) ingest particulate matter; especially the ingestion and destruction of microbes, cell debris, and other foreign matter.

Pharyngeal stage (of swallowing)   The second phase of swallowing. The bolus (mass of food) passes through the pharynx (throat) into the esophagus. An involuntary stage.

Pharynx   The throat. A tube that starts at the internal nares and runs partway down the neck. It opens into the esophagus posteriorly and the larynx anteriorly.

Phosphatase   A brush border enzyme. Breaks down nucleotides into nitrogenous bases, pentoses, and phosphates.

Phosphorus   A bulk mineral. Functions: bone development, nerve and muscle function, buffer systems, enzyme component, and energy transfer (ATP).

Phosphorylation   The addition of a phosphate group to a chemical compound. Types include substrate phosphorylation, oxidative phosphorylation (occurs in the electron transport chain), and photophosphorylation (occurs in plants).

Plaque   Two definitions. In teeth: a mass of bacterial cells, dextran (polysaccharide), and other debris that adheres to the teeth. In arteries: a cholesterol-containing mass in the tunica media (middle coat) of arteries.

Plasma   The fluid portion of blood.

Plexus   A network of nerves, veins, or lymphatic vessels.

Plexus of Auerbach   *See* Myenteric plexus.

Plexus of Meissner   *See* Submucosal plexus.

Plicae circulares   *See* Circular folds.

Polysaccharide   A carbohydrate in which three or more monosaccharides are joined chemically.

Polyunsaturated fatty acid   A fatty acid that contains more than one double covalent bond between its carbon atoms.

Abundant in triglycerides of corn oil, safflower oil, and cottonseed oil.

Postabsorptive state   Metabolic state during which absorption is complete and energy needs of the body must be satisfied. Also called the *fasting state*.

Potassium   A bulk mineral. Important for nerve and muscle function.

Premolar   *See* Bicuspid.

Preoptic area   A group of neurons in the hypothalamus that controls temperature reflexes.

Primary active transport   One of two types of active transport. The energy derived from splitting ATP *directly* moves a substance across the plasma membrane. The most prevalent primary active transport pump is the sodium pump. It pumps sodium ions (Na^+) out of the cell against their concentration gradient, maintaining a low concentration of sodium ions inside the cell (in the cytosol).

Procarboxypeptidase   The inactive form of carboxypeptidase. It is secreted by acinar cells in the pancreas and activated by the enzyme trypsin.

Proctology   The branch of medicine that treats the rectum and its disorders.

Prostaglandin   A membrane-associated lipid composed of 20-carbon fatty acids with 5 carbon atoms joined to form a cyclopentane ring. Released in small quantities and acts as a local hormone.

Protein   An organic compound consisting of carbon, hydrogen, oxygen, nitrogen, and sometimes sulfur and phosphorus. Made up of amino acids linked by peptide bonds.

Provitamin A   *See* Beta-carotene.

Pulp (of tooth)   A connective tissue containing blood vessels, nerves, and lymphatic vessels. Found in the pulp cavity of a tooth.

Pulp cavity   A cavity filled with pulp within the crown and neck of a tooth.

Pyloric antrum   *See* Antrum.

Pyloric canal   The narrow portion of the pylorus (of the stomach) that leads into the duodenum.

Pyloric sphincter   A thickened ring of smooth muscle through which the pylorus (of the stomach) communicates with the duodenum. Also called the *pyloric valve*.

Pyloric valve   *See* Pyloric sphincter.

Pylorus   The inferior region of the stomach that connects to the duodenum.

Pyridoxine   *See* Vitamin B_6.

Rectum   The last 20 cm (7 in.) of the gastrointestinal tract, from the sigmoid colon to the anus.

Regurgitate   *See* Vomit.

Retinal   A source (precursor) of vitamin A. Present in milk, cheese, fortified margarine, and butter.

Retroperitoneal organs   Organs located behind the peritoneum. The kidneys, pancreas, and some portions of the large intestine.

R group   The portion of an amino acid that distinguishes it from other types of amino acids. An amino acid consists of a central carbon atom to which four groups are attached: a hydrogen atom, an amino group, a carboxyl group and an R group. Also called the *variable group*.

Riboflavin   *See* Vitamin B_2.

Ribonuclease   A digestive enzyme found in pancreatic juice. It splits ribonucleic acid (RNA) into nucleotides.

Ribonucleic acid (RNA) A single-stranded nucleic acid constructed of nucleotides consisting of one of four possible nitrogenous bases (adenine, cytosine, guanine, or uracil), a 5-carbon sugar, and a phosphate group. There are three types, and all three are involved in protein synthesis.

Right colic flexure The portion of the transverse colon that forms a bend (flexure) under the liver. Also called the *hepatic flexure*.

Rivinus's duct *See* Sublingual gland.

Root The portion of a tooth embedded in the jaw bone.

Root canal A narrow extension of the pulp cavity lying within the root of a tooth.

Roughage *See* Fiber.

Round ligament A remnant of the umbilical vein of the fetus that is attached to the liver. Also called the *ligamentum teres*.

Rugae (singular: ruga) Large folds in the mucosa of an empty hollow organ, such as the stomach and vagina.

Saliva A clear, alkaline, somewhat viscous secretion produced mostly by the three pairs of salivary glands. Contains various salts, mucin, lysozyme, amylase, and lipase. Saliva dilutes the numbers of microbes and washes them from the surfaces of the teeth and mucous membranes of the mouth, preventing colonization.

Salivary amylase A digestive enzyme present in the saliva. Initiates the chemical breakdown of starch by reducing long-chain polysaccharides to the disaccharide maltose, the trisaccharide maltotriose, and short-chain glucose polymers called alpha-dextrins.

Salivary gland One of three pairs of glands that lie outside the mouth and pour their secretory product (called saliva) into ducts that empty into the oral cavity. Three types: parotid, submandibular, and sublingual.

Salivation The secretion of saliva.

Salivatory nuclei Two regions in the brain stem that control the secretion of saliva. The superior and inferior salivatory nuclei.

Satiety A sensation of fullness with lack of desire to eat.

Saturated fatty acid A fatty acid that contains no double bonds between any of its carbon atoms; all are single bonds and all carbon atoms are bonded to the maximum number of hydrogen atoms. Found naturally in triglycerides of animal foods such as meat, milk, milk products, and eggs.

Secondary active transport One of two types of active transport. The energy stored in ion gradients drives substances across the membrane. Since the ion gradients are established by primary active transport pumps, secondary active transport *indirectly* uses energy obtained from splitting ATP.

Secretin A hormone secreted by enteroendocrine cells in the small intestine. It stimulates the secretion of bicarbonate ions by the pancreas and liver.

Secretion Production and release from a gland cell of a fluid, especially a functionally useful product as opposed to a waste product.

Segmentation Localized contractions in portions of the small intestine containing food. It mixes chyme (partially digested food) with digestive juices.

Serosa *See* Serous membranes.

Serous membranes Membranes that line a body cavity that does not open directly to the exterior, and cover the organs that lie within the cavity. They consist of thin layers of areolar connective tissue covered by a layer of mesothelium. They are composed of two portions: the parietal portion attached to the cavity wall and the visceral portion attached to the organs inside the cavities. Include the membranes that line the pleural, pericardial, and peritoneal cavities.

Shell temperature The body's temperature at the surface (skin).

Sigmoid colon The S-shaped portion of the large intestine that begins at the level of the left iliac crest, projects inward to the midline, and terminates at the rectum at about the level of the third sacral vertebra.

Simple carbohydrates Monosaccharides (glucose, fructose, and galactose) and disaccharides (maltose, sucrose, and lactose).

Simple diffusion The mixing of ions and molecules in a solution due to their kinetic energy (random movement). The net movement of molecules from a region of high concentration to a region of low concentration.

Sinusoid A microscopic space or passageway for blood. Located in certain organs such as the liver and spleen. Also called a *sinusoidal capillary*.

Sinusoidal capillary *See* Sinusoid.

Small intestine A long tube of the gastrointestinal tract. It begins at the pyloric sphincter of the stomach, coils through the central and lower part of the abdominal cavity, and ends at the large intestine. Divided into three segments: duodenum, jejunum, and ileum.

Sodium A bulk mineral. Important for nerve and muscle function, buffer systems, and electrolyte balance.

Soft palate The posterior portion of the roof of the mouth. Extends posteriorly from the palatine bones and ends at the uvula. A muscular partition lined with mucous membrane.

Somatostatin *See* Growth hormone-inhibiting hormone.

Somatotropin *See* Human growth hormone.

Sphincter A circular muscle that acts as a valve, constricting an opening.

Sphincter of Oddi *See* Sphincter of hepatopancreatic ampulla.

Sphincter of the hepatopancreatic ampulla A circular muscle at the opening of the hepatopancreatic ampulla in the duodenum. Also called the *sphincter of Oddi* or *Oddi's sphincter*.

Splenic Pertaining to the spleen.

Splenic flexure *See* Left colic flexure.

Starch A polysaccharide consisting of hundreds or thousands of glucose units. The principal storage form for sugar in plants.

Stellate reticuloendothelial cell Phagocytic cell that lines the sinusoid (capillary) of the liver. Also called a *Kupffer cell*.

Stensen's duct *See* Parotid gland.

Stepwise A gradual progression; step by step.

Steroid A subclass of lipids. All lipids consist of four interconnected carbon rings to which polar groups may be attached.

Stool *See* Feces.

Stomach The J-shaped enlargement of the gastrointestinal tract directly under the diaphragm in the epigastric, umbilical, and left hypochondriac regions of the abdomen, between the esophagus and small intestine.

Sublingual gland One of a pair of salivary glands situated in the floor of the mouth under the mucous membrane and to the side of the lingual frenulum, with a duct (Rivinus's duct) that opens into the floor of the mouth.

Submandibular gland   One of a pair of salivary glands found beneath the base of the tongue under the mucous membrane in the posterior part of the floor of the mouth, posterior to the sublingual glands, with a duct (Wharton's duct) situated to the side of the lingual frenulum. Also called the *submaxillary gland*.

Submaxillary gland   *See* Submandibular gland.

Submucosa   A layer of connective tissue located beneath a mucous membrane. The submucosa connects the mucous membrane (mucosa) to the muscularis layer.

Submucosal plexus   A network of autonomic nerve fibers located in the submucosa of the small intestine. Also called the *plexus of Meissner*.

Substrate   A substance (metabolite) with which an enzyme reacts.

Substrate phosphorylation   A high-energy phosphate group is transferred directly from a substrate (metabolite) to a molecule of ADP, forming ATP.

Sucrase   A brush border enzyme. It breaks the disaccharide sucrose into fructose and glucose.

Sucrose   A disaccharide (double sugar) consisting of fructose and glucose.

Sulfur   A bulk mineral. Functions: hormone component; vitamin component; and ATP production.

Taenia coli   (plural: taeniae coli)   One of three flat bands of thickened, longitudinal muscles running the length of the large intestine.

Testosterone   A male sex hormone (androgen). It controls body growth, the growth and development of male sex organs, the development and maintenance of secondary sex characteristics, and the production of spermatozoa. It also increases the metabolic rate.

Thiamine   *See* Vitamin B_1.

Thyroxine (T_4)   Thyroxine (T_4) is a hormone secreted by the thyroid gland. It regulates organic metabolism, growth and development, and the activity of the nervous system.

Tocopherols   *See* Vitamin E.

Tongue   A large skeletal muscle covered by a mucous membrane located on the floor of the oral cavity.

Tonic contraction   Partial contraction of a muscle due to a steady input of nerve impulses. Results in muscle tone.

Trace minerals   13 minerals that are required in extremely small quantities, but are essential. They function as components of enzymes, hormones, and other molecules in the body. They include copper, cobalt, chromium, fluorine, iodine, iron, manganese, molybdenum, selenium, silicon, tin, vanadium, and zinc.

Transamination   Formation of a new amino acid. For example, the transfer of an amino group to a keto acid forms a new amino acid.

Transverse colon   The portion of the large intestine extending across the abdomen from the right colic (hepatic) flexure to the left colic (splenic) flexure.

Triacylglycerol   *See* Triglyceride.

Tricarboxylic acid cycle (TCA)   *See* Krebs cycle.

Triglyceride   A lipid formed from one molecule of glycerol and three molecules of fatty acids. The body's most highly concentrated source of energy. Also called a *neutral fat* or *triacylglycerol*.

Trypsin   A digestive enzyme secreted by acinar cells of the pancreas in the inactive form called trypsinogen. Activated by the enzyme enterokinase. It breaks down proteins into fragments called peptides.

Trypsinogen   The inactive form of the protein-digesting enzyme trypsin.

Upper esophageal sphincter   A sphincter muscle (valve) at the upper end of the esophagus.

U.S. RDA   *See* U.S. Recommended Daily Allowance.

U.S. Recommended Daily Allowance   The quantity of a specific vitamin or mineral that should be ingested daily.

Uvula   A soft, fleshy mass; especially the V-shaped structure hanging down from the soft palate in the posterior portion of the oral cavity.

Variable group   *See* R group.

Vermiform appendix   A twisted, coiled tube attached to the cecum (first portion of the ascending colon). Also called the *appendix*.

Vermilion   The transition zone of the lips, where the outside skin meets the inside mucous membrane. The color of the blood in the underlying blood vessels is visible through the transparent surface of the vermilion.

Very low-density lipoproteins (VLDLs)   Lipoproteins that transport triglycerides synthesized by liver cells to adipose cells (fat cells), where VLDLs are converted into LDLs.

Vestibule   A small space or cavity at the beginning of a canal. The mouth, inner ear, larynx, nose, and vagina have vestibules.

Villus   (plural: villi)   A projection of the lining of the small intestine. Contains connective tissue, blood vessels, and a lymphatic capillary (lacteal). Increases the surface area, which increases the efficiency of absorption of digested foods.

Viscera   (singular: viscus)   The organs inside the ventral body cavity.

Visceral   Pertaining to the organs or to the covering of an organ.

Visceral peritoneum   Serous membrane that covers some of the abdominal organs (viscera).

Vitamin   An organic molecule that is not synthesized in the body in sufficient amounts to maintain health. Must be included in the daily diet. Acts as a catalyst in metabolic processes in the body.

Vitamin A   Fat-soluble vitamin. Functions: constituent of visual pigment; maintenance of epithelium; mucopolysaccharide synthesis.

Vitamin B_1   Water-soluble vitamin. Component of coenzymes involved in the formation of acetyl coenzyme A. Also called *thiamine*.

Vitamin B_2   Water-soluble vitamin. Component of the coenzyme FAD (involved in ATP production). Also called *riboflavin*.

Vitamin B_6   Water-soluble vitamin. Coenzyme involved in amino acid and lipid metabolism. Also called *pyridoxine*.

Vitamin B_{12}   Water-soluble vitamin. Coenzyme involved in nucleic acid metabolism. Also called *cobalamin*.

Vitamin C   Water-soluble vitamin. Functions: coenzyme involved in collagen synthesis; maintains intercellular matrix of cartilage, bone, and dentin. Also called *ascorbic* *acid.*

Vitamin D   Fat-soluble vitamin. Functions: promotes growth and mineralization of bones; increases absorption of calcium in GI tract.

Vitamin E   Fat-soluble vitaimin. Functions: antioxidant to prevent cell membrane damage; prevents breakdown of vitamin A and fatty acids. Also called *tocopherols*.

Vitamin K   Fat-soluble vitamin. Essential for liver synthesis of prothrombin and other clotting factors.

Voice box   *See* Larynx.

VLDL   *See* Very low-density lipoprotein.

Voluntary stage   (of swallowing)   The first stage of swallowing. A bolus (food mass) is forced to the back of the oral cavity and into the oropharynx by movement of the tongue upward and backward against the hard palate.

Vomit   To eject the contents of the stomach through the mouth. Also called *regurgitate*.

Water-soluble vitamins   Vitamins B_1, B_2, B_6, and B_{12}; niacin; folacin; biotin; pantothenic acid; and vitamin C.

Wharton's duct   *See* Submandibular gland.

Wisdom teeth   3rd molars.

Zymogenic cell   *See* Chief cell.

Bibliography

Curtis, Helena.   *Biology,* 3rd ed.
New York : Worth, 1979.

Dorland, William Alexander.   *Dorland's Illustrated Medical Dictionary,* 27th ed.
Philadelphia : W. B. Saunders, 1988.

Ganong, William F.   *Review of Medical Physiology*, 15th ed.
Norwalk, Connecticut : Appleton & Lange, 1991.

Junqueira, L. Carlos, Jose Carneiro, and Robert O. Kelley.   *Basic Histology*, 6th ed.
Norwalk, Connecticut : Appleton & Lange, 1989.

Kimball, John W.   *Biology*, 4th ed.
Reading, Massachusetts : Addison-Wesley, 1978.

Melloni, B.J., Ida Dox, and Gilbert Eisner.   *Melloni's Illustrated Medical Dictionary*, 2nd ed.
Baltimore : Williams & Wilkins, 1992.

Tortora, Gerard J. and Sandra Reynolds Grabowski.   *Principles of Anatomy and Physiology,* 7th ed.
New York : HarperCollins, 1993.

Vander, Arthur J., James H. Sherman, and Dorothy S. Luciano.   *Human Physiology,* 5th ed.
New York : McGraw-Hill, 1990.

Woteki, Catherine E. and Paul R. Thomas, eds.   *Eat For Life,* 1st ed.
New York : HarperCollins, 1993.